MEASURE
· FOR ·
MEASURE

THE JOHNS HOPKINS UNIVERSITY PRESS

MEASURE
· FOR ·
MEASURE

The Story of Imperial, Metric, and Other Units

ALEX HEBRA

The Johns Hopkins University Press
Baltimore and London

The Johns Hopkins University Press
2715 North Charles Street
Baltimore, Maryland 21218-4363
www.press.jhu.edu

Library of Congress Cataloging-in-Publication Data

Hebra, Alex, 1919–
    Measure for measure : the story of imperial, metric, and other
units / Alex Hebra.
        p. cm.
    Includes bibliographical references and index.
ISBN 0-8018-7072-0 (hardcover : alk. paper)
    1. Weights and measures—History.   I. Title.
    QC83 .H43 2002
    389'.1'09—dc21   2002001840

A catalog record for this book is available from the British Library.

To
Dr. Heinrich Mache,
*Professor Ordinarius at the Technical University
of Vienna,*

who taught me the love of physics

· Contents ·

· Preface ·

When the Puritans landed at Plymouth Rock, they brought English culture with them—including the use of the foot and the pound to measure lengths and weights, respectively. Some three hundred years later, Americans still use these units of measure. Had the French and Indian Wars ended differently—if Montcalm had defeated Wolfe on the Plains of Abraham—we might have become instead a French-speaking country—and a metric one.

As it is, we are very much in the minority. With the exception of the United States, North and South Yemen, Burma, and Brunei, all countries of the world have in time "gone metric" by adopting the International System of Weights and Measures (SI), proposed in 1960 for worldwide use. Most people are surprised to learn that the United States had already become an "officially metric nation" in 1893, the year the secretary of the U.S. Treasury declared a set of metric standards as the country's "fundamental standards of length and mass." But metric units were never able to gain nationwide acceptance. Throughout the twentieth century, a battle was waged between the yard and the meter, the kilogram and the pound, and between Fahrenheit and Celsius.

One significant step toward metrication was taken after World War II, with the rounding down of the inch to 25.4 millimeters, which made the length of 5 inches equal to exactly 127 millimeters. Since then, however, the introduction of metric units in the United States has remained on the back burner.

Although the original signatories of the Treaty of the Meter of 1875 are quite comfortable with their new system, the latecomers have been ambivalent. The English daily newspaper *The Guardian*, for example, reported that thirty-three years after the adoption of the metric system of weights and measures, 74 percent of people

surveyed still preferred the "old ways of measurement," and a mere 19 percent found the metric system more convenient. A "bitter resentment against the government and European Union" was in the forecast when the use of metric units would become "compulsory and enforceable by criminal sanctions."

Sure enough, while the rest of the world needlessly worried about the Y2K problem, that measure, planned for January 1, 2000, was without much ado, set back for another decade.

If such "bitter resentment" flares up in a country where the expense of introducing new units has already been budgeted, convincing Americans to finance such a change is likely to be a still tougher job. Adoption of new units would make millions of standard parts and equipment in American industry obsolete. A wealth of molds, dies, stamps, templates, fixtures, and other manufacturing necessities would end up on the scrap heap. In short, the entire tooling for serial production of standardized components, from screws and bolts to gears and sprockets, would have to be replaced. Retooling in metrics would cost American industries tens if not hundreds of billions of dollars, an outlay that would ultimately be passed on to consumers.

This book looks at some of the many ways that humans have devised to measure everything from the length of a strip of cloth, to the brightness of a candle and the loudness of a rock concert. The diverse systems of measures we have invented have both helped and hindered scientific progress. Having standards of measurement for one's local merchants or scientists helps to foster fair trading, enables the comparison of scientific experiments, and hastens the development of local technologies. But measurements can be counterproductive unless the system employed becomes universal. The recent crash of a NASA probe to Mars, for example, was due to errors caused by instructions that were both metric and imperial.

In showing how various measurement systems have evolved, I have opted to add some colorful stories from the realms of both myth and history, and to explain the physics behind such little known yet intriguing devices as the Jefferson pendulum and the grease-spot photometer. I have also included instructions on how to build a hydroelectric dam on another planet, though "hydro"

might be misleading because water may not exist there. I have added these asides because, like the motto of the British Broadcasting Corporation, I want to "inform, educate, and entertain."

When you finish this book, which is, by the way, printed in a decidedly nonmetric $5\frac{1}{2} \times 8\frac{1}{2}$ inch format, I hope that you will not only have enjoyed reading it, but will have learned something new about the universe in which we live and will understand more about the units scientists use to describe our world. And should your vacation plans include the metricated parts of the world, you will go prepared and will have found in this book the proverbial that an "ounce of prevention is worth . . ."—oops, old habits are hard to break—make that "28 to 29 grams of prevention are worth 0.454 kilos of cure."

· Acknowledgments ·

At a time that was more idealistic than ours, the French philosopher René Descartes theorized on the infinite perfection of the divine, which led some of his students to conceiving the harmonic functioning of the world we live in as the consequence of a kind of sublime optimization of all its constituents. In the final analysis, your writing pen had to be the best—under the given circumstances—of all possible pens. Likewise, your mind must be the best of all possible minds, the emperor the best of all possible emperors, and your mother-in-law . . . well, perhaps we shouldn't linger about that.

If such notions have not endured, it might have to do with contemporary people's inability to view their jobs as the best of all possible jobs, and their bosses—OK, you get the idea. However, I can personally witness a case where this philosophical concept still obtains: Dr. Trevor Lipscombe, editor-in-chief at the Johns Hopkins University Press, who was for me and this book undoubtedly the best of all possible editors within the 15-billion light-year radius of our universe. I thank him for joining his competence and professionalism to his unlimited energy in efforts to enhance the book's structure and presentation, and to enrich it with a host of new ideas and suggestions. Without his guidance, this work would never have evolved into the present book.

Likewise, my gratitude goes to Alice Calaprice, author of two books on Albert Einstein, one of which has been translated in twenty-two languages, for lending her creativity and outstanding linguistic competence to guarantee all aspects of quality to this work.

Further, I give my highest appreciation for the comprehensive marketing program that Mary Katherine Callaway, of The Johns Hopkins University Press, has prepared to maximize the book's availability and to direct it into the proper channels that promise highest impact and effective distribution.

My sincerest thanks go to the institutions and people whose support made my work possible, principally the National Institute of Standards and Technology, particularly Karma Beal, for supplying me with illustrations and a host of fundamental information on units and standardization.

I equally thank Kristina Reichenbach, Dr. Geuther, and Dr. Harlieb at the DIN (Deutsches Institut für Normung e.V.), Berlin, for literature and personal information on DIN and ISO (International Organization for Standardization) standards and the status of related efforts being made abroad.

My special thanks go to Prof. Hannspeter Winter, Technische Universität Wien, Institut für Allgemeine Physik, Vienna, Austria, for his interesting comments, suggestions, and encouragement, and to Dr. Andrew Digby of the Royal Observatory of Edinburgh, U.K., for working out information on brightness and frequency of fixed stars that I could not have obtained otherwise.

Thanks also to Dr. Gareth Ruddiger for helping me through Greek mythology with his extraordinary knowledge of ancient history, and Kevin Thorp, Visual Dynamics, for the ideas and illustrations he contributed and for his indefatigable advice on computer editing and software problems.

Thanks likewise to Sheila Amos, Mt. Pleasant Branch of the Charleston County Library, for skillfully obtaining badly needed information that was hard to come by, and to Brian Dengel, Product Group manager at Quality Transmission Components, Division of Designatronics, Inc., for contributing valuable engineering drawings on gearing standards.

In old and unwavering friendship, I thank Jim Wilson, chief engineer at the Gates Rubber Company, for allowing me access to his private collection of insider's literature on aviation.

Last but not least, I give my loving thanks to my son Alexandre for taking care of my computer system and for solving a long line of software and hardware problems, and to my wife, Gerda, and my other children, Andre and Renata, who must have had a hard time putting up with my obsession of preparing this work and whose patience and encouragement I cannot praise highly enough.

MEASURE
· FOR ·
MEASURE

· 1 ·

Blessing Our Countings

Units and Numbers

There comes a great moment in a young child's life when he or she learns how to count. Immediately, the child wants to count everything—toys, fruit, the people at a party—and does so with a sense of glee. This may not seem important—after all, everyone learns how to count eventually. But a decisive intellectual development has taken place. The youngster can conceptualize several different objects as perfectly identical. Red, square-shaped blocks and cylindrical blue blocks are both blocks and can be counted together. *In short, the capacity to count implies the intellectual power of abstraction.*

The precondition for counting is to categorize. If we rigorously followed the adage "You can't add apples and oranges," we could not count at all, for there are no two perfectly identical objects in this world. But if we can't add apples and oranges, could we add other—if ever so slightly different—objects? The answer is that prior to counting, we must specify what we intend to count. The apple/orange proverb is perfectly correct as long as we count either apples *or* oranges. But it does not apply to counting fruit, edibles, natural products, or any other assortment that incorporates both apples and oranges. So, all our accounts depend on the kind of categories we define. The statement "Ten pieces of fruit are in that basket" means the basket may contain six oranges and four apples; or two prunes, five pears, and three peaches; or whatever other combination there may be, since it refers to "fruit"

rather than individual varieties. Likewise, speaking of "five fingers on each hand" implies that we ignore the substantial differences between thumb, index finger, and the others.

To the Bare Bones

We cannot know for certain and can only speculate about the ways people counted—and maybe calculated—in the era before they left written accounts, which began to appear around 30,000 B.C. At that time, Stone Age people had devised a system of tally by groups. A 7-inch-long shin bone of a young wolf, discovered in Czechoslovakia in 1937, was found engraved with fifty-five deeply cut notches, arranged in groups of five, proving that even Stone Agers had a rudimentary notion of grouping figures by their orders of magnitude. It is interesting to note that our words for counting and writing have much to do with the early tools used to perform them: the English word "write" comes from the Anglo-Saxon "writan," meaning to scratch; and our word "calculate" goes back to the Latin term "calculus," meaning pebble.

Tally sticks are among the oldest means by which to count. As late as the nineteenth century, the "double tally stick" was still used by the Bank of England to register the amounts of their clients' deposits. Once it was engraved with the appropriate number of marks, the stick was cut into two halves: one called the "foil," which the bank kept; and the other the "stock," for the customer, who in the process became the "stockholder" owning "bank stock." For withdrawals, the stock was checked against the foil, and if they agreed, the respective amount of currency was paid with a written certificate that later became a "check." It took an act of Parliament in 1846 to abolish tally sticks, which had been around since the twelfth century, in the days of Richard the Lionheart.

As early as 3500 B.C., the Egyptians had a base-10 system of hieroglyphic numerals (Fig. 1.1), up to hundreds of thousands and even millions. Hieroglyphs underwent some modifications from the Old Kingdom (2700 B.C. to 2200 B.C.) to the Middle Kingdom (2100 B.C. to 1700 B.C.), to the New Kingdom (1600 B.C. to 1000 B.C.), but retained their broadly similar style. The 3000 B.C. hieroglyphs on Narmer or Menes, the first of their kind in upper and

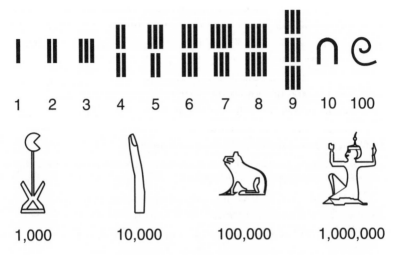

Figure 1.1 Egyptian numerals

lower Egypt, document the existence of thousands of heads of cattle and a similar abundance of enslaved prisoners of war.

The numbers from 1 to 9 were shown by the appropriate total strokes, 10 by a hobble for cattle, 100 by a coil of rope, 1000 by the drawing of a lotus plant, 10,000 by a finger, 100,000 by a tadpole or frog, and a million by a divine figure, arms raised. What we have here is an "additive number system" because the numerical value of a certain hieroglyph does not depend on its position within the number. Additive number systems were also used by the Sumerians, who lived from 2900 through 1800 B.C. in the area now encompassing Kuwait and northern Saudi Arabia, and were also present in ancient Greece and Rome. Although Egyptians had a symbol for zero, it did not become part of their numbering system. Numbers were combined by writing the higher numeral successively in front of the lower one.

In "positional number systems," including our decimal system, the numerals for denoting a number take different place values according to their position. For instance, the number 1 in decimal stands for 1, 10, 100, etc., depending on its placement as the first, second, or third digit. This principle reaches back to

cuneiform (from the Latin *cuneus,* "wedge," which, for the experts, is a second-declension masculine noun) inscriptions found on Babylonian clay tablets of around 1900 B.C., which had been engraved with a hollow stylus while the clay was still pliant. Drying and subsequently heating the tablets close to 1000°C weatherproofed them enough to allow them to last for millennia.

Ancient Babylon was inhabited from the time of the first cities (4500 B.C.) until well after the birth of Christ, spanning an area in modern Iraq then called Mesopotamia, meaning "between the rivers" in Greek. Irrigated by the rivers Tigris and Euphrates, the country prospered and built many great trading cities such as Ur and Babylon. The Babylonian number system (Fig. 1.2) uses wedges for the numbers from 1 to 9, and hooks for multiples of 10; but it lacks the zero, so that a unique symbol for 10 becomes necessary.

Armed with this set of basic symbols, Babylonians could have assembled all numbers up to 99; but for reasons not easily understandable, they stopped short at 59. Therefore, their system is sexagesimal, or base 60, though a truly sexagesimal number system would need unique symbols for the first fifty-nine numbers. To be

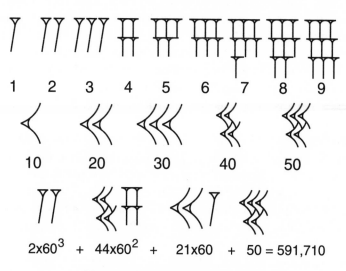

$$2 \times 60^3 \ + \ 44 \times 60^2 \ + \ 21 \times 60 \ + \ 50 = 591,710$$

Figure 1.2   Babylonian numerals

precise, we should talk of a "decimal-embedded sexagesimal system." Nevertheless, remnants of the 60-based system from ancient Babylon are still with us as subunits of time and angles: minutes, seconds of arc, and the hour. Incidentally, the Mayans, who used human heads (and parts thereof) to represent numbers, also used a positional number system.

Babylonian mathematics flourished with its particular set of numbers, even if it was sexagesimal. For instance, the inscription on a tablet housed in the British Museum shows the Babylonians' knowledge of Pythagoras's theorem. Translated into English, it reads: "4 is the length, and 5 the diagonal. What is the breadth? Its size is not known. 4 times 4 is 16. 5 times 5 is 25. You take 16 from 25 and there remains 9. What times what shall I take in order to get 9? 3 times 3 is 9, 3 is the breadth." Today called the "three/four/five" triangle, it has maintained its place as a handy surveyor's tool for those who forgot the laser pointer at home or the theodolite back in camp. Proof of this theorem has fascinated mathematicians throughout the centuries. For example, James Garfield, while serving in the House of Representatives on the way to becoming the twentieth president of the United States, discovered in 1876 a version of a proof based on the formula for the area of a trapezoid; it is still used in most high school textbooks. The multitalented Garfield, by his own account, also amused his friends by simultaneously writing Latin with one hand and Greek with the other.

In ancient Greece, Archimedes, scientist and philosopher, designed his very own multiplicative number system to solve far-reaching problems, such as finding the volume of the universe, which he reports equivalent to the volume of 10^{64} grains of sand. Setting the size of the grain of sand at one millimeter (about $\frac{5}{128}$ inch), the radius of a sphere holding 10^{64} such grains comes to 141 light-years, approximately the distance to the V-shaped Hyades star cluster in the constellation of Taurus the Bull. Although Archimedes' guess falls far short of modern figures, his universe would still house most of the brightest stars we see in the sky, including Sirius, Vega, Arcturus, and of course Alpha Centauri, our closest stellar neighbor apart from the sun. Archimedes' understanding of cosmic distances, unmatched for over two millennia, is an incredible achievement.

For the more earthbound applications, like counting money and weights, Greeks of the first millenium B.C. used the "acrophonic" number system, which associates basic numbers to the first letters of their names (Fig. 1.3): P from *pente* for 5, D from *deka* for 10, H from *hekaton* for 100, X from the Greek spelling of *khilioi* for 1000, and M from *murioi* for 10,000. The exception is the 1, which was a vertical scratch (|).

Later in their history, Greeks adopted the "alphabetic" number system, where the letters stood for numbers according to their position in the twenty-four letters of the Greek alphabet of those times. Thus the letters α through θ were the numbers 1 to 9, the letters ι to koppa stood for 10 through 90, and the letters ρ to san for 100 through 900. Hence, the obsolete letters "digamma" (6), "koppa" (90), and "san" (900) were made part of the alphabet of that time. Combined, these letters formed an additive number system: for instance,: ια = 11, ιβ = 12, ιθ = 19, ρμζ = 147.

Though the year A.D. 876 marks the first use of the zero in India, whence our Indo-Arabic numbers stem, the Maya of Mexico and Peru had a zero sign already centuries earlier. Their 20-based system (Fig. 1.4), consists of dots (= 1) and bars (= 5). Numbers were combined from the bottom up, the least significant of the group below the rest. Addition in this system consists of adding dots and bars, yet keeping in mind that a bar must replace a row of five dots. Those familiar with Welsh or French should be no strangers to vigesimal (20-based) systems, which in principle resemble the Mayan system. We also know that the English-speaking world uses some base-20 expressions, such as in Lin-

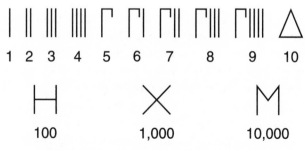

Figure 1.3   Grecian acrophonic numbers

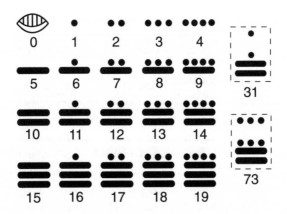

Figure 1.4 Mayan numerals

coln's famous address at Gettysburg, which begins "Four score and seven years ago."

In ancient Rome, the numbers 1, 10, 100, and 1000 were expressed by the letters I, X, C, and M. To save space and writing material, Romans introduced the additional letters V, L, and D for 5, 50, and 500. Unlike most other, basically additive systems of numbers, the Roman system includes subtraction. The number nine, for example, reads IX, forty reads XL, and so on. In Roman numerals, the Declaration of Independence occurred in MDC-CLXXVI and contains addition only, but the date of Christopher Columbus's discovery of the Americas, MCDXCII, uses subtraction twice. Subtraction kept Roman numbers from becoming chains of unsuitable length, but at the same time it precluded such handy algorithms as we use for multiplication and division. After all, it's harder to find a suitable algorithm for performing the division LXXII ÷ IX = VIII than for 72/9 = 8. Though they have fallen out of favor today, Roman numerals are traditionally used to number "front matter" (table of contents, preface, etc.) in books, and, in some countries, to denote the month in dates. In this system, the first world war ended in 11.XI.1918.

Our current positional number system, in which the value of each digit is determined by its position within the number, began around A.D. 600. It stemmed from India's Brahmi-number

characters, supplemented by the introduction of a zero, and was augmented in the tenth century A.D. by the introduction into the West of the Arabic number symbols in use today. Several people have been credited with this, one of them being Gerbert d'Aurillac, a mathematician better known to historians as Pope Sylvester II, the only pope to have had his mathematical works published. Leonardo of Pisa, also known as Leonardo Fibonacci, an Italian mathematician in the court of Emperor Friedrich II von Hohenstaufen, wrote the *Liber Abaci* in 1202. It was the first systematic introduction into the Indian number system but also contained letters on number theory and geometry. The *Liber Abaci* also introduced the famous problem of the reproducing rabbits. To solve it, Fibonacci introduced a sequence of numbers where each successive term becomes the sum of the two preceding terms. Making the first two terms of the series 1 and 1, the term next in line becomes 1 + 1 = 2, followed by 1 + 2 = 3. This way, one can generate strange-looking series such as

$$1, 1, 2, 3, 5, 8, 13, 21, 34, \text{etc.}$$

By changing the starting numbers (from 1,1 to 3,6, say), you get a new sequence. Eight hundred years after Fibonacci numbers first appeared, new books and a journal are still being devoted to their mathematical properties.

In A.D. 1299, the city of Florence issued an ordinance prohibiting the use of Hindu-Arabic numerals because they were easier to falsify than Roman numerals. This and similar decrees elsewhere delayed the acceptance of Arabic numerals throughout Europe until the fifteenth century.

In terms of number systems, we have a new champion today: the binary. It is fair to say that in the short time it has been used so far, many more mathematical operations have been performed in binary than in all other number systems combined through history. This is so because human brains were needed to operate in all those 5, 10, 20, and 60-based systems, while the central processing units (CPU) at the heart of our computers can handle binary operations in record time, far faster than humanly possible. While mechanical

calculators, including the Chinese abacus, emulated human ways of handling the basic arithmetical operations of adding, subtracting, multiplying, and dividing, computers must make do with the two options an electrical switch has to offer: on and off. These options made the binary system a natural in computer engineering.

Until this time, the binary system had been the stepchild in the world of numbers. Occasionally, a well-meaning citizen suggested a binary-scaled currency because it would allow for adding up any desired amount with only one piece of each value. For instance, if coins of 1, 2, 4, 8, 16, 32, and 64 cents were issued, one could combine them to 16 + 4 = 20 cents or 32 + 8 = 40 cents; or one could have quantities like 57 = 32 + 16 + 8 + 1 or \$1.13 = 64 + 32 + 16 + 1, etc. Such a system was expected to cut down on the total of coins to be issued, but it would have unduly complicated the process of getting change.

Thus, we view the binary division of the dollar as utopian, and yet, the old monetary system in the United Kingdom had its farthings, worth one quarter of a penny ($\frac{1}{4}$ d), the ha'penny ($\frac{1}{2}$ d), pennies (1 d), shillings (12 d). Twenty shillings made £ 1, and 21 shillings made a guinea. Lawyers and doctors billed in guineas.

In a different binary setting, the "German Treatise" of the sixteenth century mandated the length unit of the rood (equal to $16 = 2^4$ feet). This was defined as the space taken by "sixteen men of all sizes" standing in line with their left feet toe to heel—an interesting attempt at creating a reproducible standard of length by averaging out individual differences. Since then, the foot has been defined by its relation to metric standards, but the binary-based 16 ounces in a pound are still used every day.

The popularity of such units amid a decimal world rests on daily-life activities. A tablecloth, sheet, or blanket can be neatly folded into halves, quarters, or eighths by joining the edges. Letter-size sheets of paper are folded twice to fit into standard envelopes unless they are used with No. 9 regular envelopes of dimension $3\frac{7}{8} \times 8\frac{7}{8}$ inches, for which a letter-size sheet of $8\frac{1}{2} \times 11$ inches needs to be folded into three parts, something we do with a degree of precision unattainable with decimal partitions such as 5 or 10.

If you want to fold things in halves, thirds, or quarters, it's good to have a 12-based unit, as 2, 3, and 4 divide it evenly; hence, the inch and the foot. Ease of divisibility is essential in shop work, which explains the popularity of the yardstick over the meter. If we had a choice, the number 12, divisible by 2, 3, 4, and 6, might be a better selection than the 10 for our number system, though by the same token, 16 would be a valid contestant. Mind you, Ann Boleyn, the second wife of Henry VIII and the first to be beheaded, had six fingers on one hand. If the rest of us mortals had fingers in such abundance, we all would probably count in duodecimal!

A comparison of the reciprocals of the first twelve numbers ($\frac{1}{1}, \frac{1}{2}, \frac{1}{3}, \frac{1}{4}$, etc.) in duodecimal and in decimal shows why division of numbers would be easier in the duodecimal system. But we must remember that the number 12 reads "10" in the duodecimal (base 12) system, which calls for the introduction of two extra one-digit figures: A for 10 and B for 11. Further, in the division of duodecimal numbers, the first figure following the dot— now the "duodecimal point"—stands for $\frac{1}{12}$ (rather than $\frac{1}{10}$), the second-place figure for $\frac{1}{12^2} = \frac{1}{144}$, and the third-place figure for $\frac{1}{12^3} = \frac{1}{1728}$, etc. Thus, the reciprocals in duodecimal are computed as follows:

$$\frac{1}{2} = \frac{6}{12}, \text{notation } 0.6; \quad \frac{10}{3} = \frac{40}{12}, \text{notation } 0.4;$$

$$\frac{1}{8} = \frac{3}{24} = \frac{3}{2\times12} = \frac{18}{12\times12} = \frac{12+6}{12\times12}$$

$$= \frac{1}{12} + \frac{6}{144}, \text{notation } 0.16; \text{etc.}$$

Proceeding like this, we find the reciprocals of the twelve first numbers in duodecimal:

1	0.6	0.4	0.36	0.24972	0.2
0.186A3	0.16	0.14	0.12497	0.11111	0.1

The same series in decimal consists of the following:

1	0.5	0.333333	0.25	0.2	0.166666
0.142857	0.125	0.111111	0.1	0.090909	0.083333

Discounting the shaded cells, which contain divisions that do not terminate, and the trivial $\frac{1}{1} = 1$ division, we find that five reciprocals terminate in the decimal sequence compared to seven in the duodecimal series. In a 12-based number system, we thus expect an average of 40% more finite divisions than in ours.

In the days when people used paper and pencil, this system would have saved considerable amounts of brainpower. But nowadays, electronic calculators and computers have overcome such problems. These machines would not be speeded up by duodecimal input because they operate in binary code—base 2.

Numbers and . . . Numbers

Opposed to the cardinal numbers 0, 1, 2, 3, etc., which are direct descendants of piece-by-piece counting and therefore are always integers, rational numbers are fractions of cardinal numbers, for instance: $\frac{1}{2}, \frac{3}{4}, \frac{7}{10}, \frac{37}{23}, \frac{537}{71}$, etc.

Because the spacing between rational numbers can be made as small as desired, they were originally thought capable of expressing any given quantity. For instance, the difference between $\frac{37}{23} = 1.60870$ and $\frac{38}{23} = 1.65217$ is 0.04347. For a smaller step, we could use three-digit integers and convert $\frac{37}{23}$ into $\frac{370}{230}$, so that the next closest rational number becomes $\frac{371}{230} = 1.61304$. That brings the gap to $1.61304 - 1.60870 = 0.00434$. Continuing this way, we may narrow the spacing as much as we please if we use four, five, or even more digits.

In general terms, the closest rational number, x_2, to a given number, $x_1 = \frac{a}{b}$, is derived from the identity $a/b = (n \times a)/(n \times b)$, where n can be made as high as we wish. This gives for the next higher rational number, $x_2 = (n \times a + 1)/(n \times b) = a/b + 1/(n \times b) = x_1 + 1/(n \times b)$, and for the step between successive numbers, $x_2 - x_1 = 1/(n \times b)$. The higher an n we pick, the narrower the step

from one number to the next—a relation that, when n tends toward infinite, converges toward zero spacing.

This kind of reasoning led early mathematicians to visualize rational numbers as a continuous band, one that included quantities of any magnitude whatsoever. Yet they were shown to be in error when the solutions of certain quite basic problems of mathematics and geometry could not be expressed in fractions of integers. Something else was needed: a class of numbers that could not be expressed as finite fractions. The irrationals were born.

Irrational Numbers . . . the Split

Irrational numbers are terms like $\sqrt{3}$, $\sqrt[3]{2}$, and, perhaps most famously, π, familiar from the formulas for the circle's circumference $C = 2\pi R$ and area $A = \pi R^2$. A long-standing problem was the mathematical quest for a means to construct a square that was equal in area to a given circle, using only a ruler and compass as tools. This was called "quadrature of the circle," or "squaring the circle." Much effort and brainpower were spent on attempts to solve this problem. Finally, in 1770, Johann H. Lambert, the eighteenth-century physicist and the most notorious representative of German rationalism prior to Kant, presented mathematical proof that it is impossible to describe π by a fraction of integers. Thus π was irrational, and the circle could not be squared.

As π is irrational, the best our forebears could do was to come up with approximations. π was considered equal to 3 in ancient Babylon, in the Bible (I Kings 7:23–26), and as recently as 1897 by the legislators of the state of Indiana, who set out to write this value into law. Because $\pi = 3$ would simplify calculations related to circles, this law was expected to save the state's citizens time, brainpower, and tax money. Luckily, it was blocked by a contemporary professor of mathematics, who argued that the laws of nature could not be bent the way we bend laws of human origin.

Early Egyptian papyri propose for the area A of a circle of diameter d the formula $A = \left(\frac{8}{9} d\right)^2$. Compared with the correct formula $A = \frac{\pi}{4} d^2$, this leads to $\pi = \left(\frac{8}{9}\right)^2 \times 4 = 3.160$, within 0.6% of the correct figure of 3.14159. . . Modern schoolchildren often use 22/7, which is 3.14285. . . which is only 0.04% off.

While formulas for the circumference and area of regular polygons had been successfully derived by means of basic algebra, these approaches never led to an algebraic expression for π, although the method of approximating the perimeter of the circle by multisided polygons allowed Ludolph van Ceulen, in 1600, to calculate π to thirty-five decimal places.

But there is a silver lining: If we can't determine the exact area of a particular circle, we can still find it for certain geometrical figures encompassed by circles, such as the "Hippocratic Crescents" (Fig. 1.5). At first sight, one would expect all the old problems with π to resurface as we try our skills on the areas of the "two crescents subtended by intersecting the two circles centered at the midpoints of the short sides of a rectangular triangle with the circle involving the hypotenuse." Not so! Their combined areas equal the area of the triangle. Nothing complex, nothing irrational. Just base times height divided by two.

We can calculate the value of π to any desired number of decimals by use of an infinite, converging series of numbers. The concept of mathematical series is easy to understand; so easy, that Carl Friedrich Gauss (1777–1855) used it intuitively while attending elementary school in his hometown of Braunschweig. His teacher, pressed for time to correct a pile of homework, ordered the class to sum up the numbers from one to one hundred, hoping to keep

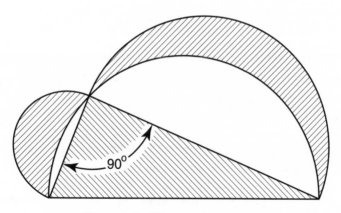

Figure 1.5 Hippocratic crescents

the class busy and quiet while he worked. But few minutes had barely passed when one of the tykes shot up his hand and shouted something equivalent to our "Gotcha!" And before the disbelieving teacher could reach for the dreaded stick, the boy continued: "Five thousand and fifty."

Of course, little Gauss had to explain his record speed addition on the blackboard, if for no other reason than to prove that he hadn't known that figure beforehand. Here is what he did. Write the numbers $1 + 2 + 3 + 4 + \ldots + 100$. Now reverse the order: $100 + 99 + 98 + 97 + \ldots + 1$, and add them up to $101 + 101 + 101 + \ldots$. Twice the sum is 100 lots of 101, which makes the solution 50×101, or 5050. Sheer genius.

What Gauss had done was to anticipate the formula $s_n = n \times (a_1 + a_n)/2$ for the sum s_n of a series of n regularly spaced numbers, the first and the last being a_1 and a_n.

While the sum of this so-called arithmetic series gets bigger with the number of terms, this is not the case with other types of series. Some converge so readily that even the sum of an infinite number of terms remains finite. The Swiss mathematician Leonhard Euler, in 1736, tried his expertise on a far more complex problem than the one Gauss tackled in school: the sum of an infinite "harmonic progression," the series of the squares of the reciprocals of the natural numbers, which, surprisingly, turned out to contain the famous term for π:

$$S = \frac{1}{1^2} + \frac{1}{2^2} + \frac{1}{3^2} + \ldots \frac{1}{n^2} \ldots = \sum_{1}^{\infty} \frac{1}{n^2} = \frac{\pi^2}{6}.$$

In principle, this expression could be used to figure the value of π, but in practice it is far too slow to be useful. The first one hundred terms (generated by computer) give $\sqrt{6} \times S_{100} = \sqrt{6} \times 1.634984 = 3.132$, far from the correct value of 3.14159 and only marginally better than that used by the ancient Egyptians. Even if it were continued to 10,000 terms, the value of π from that series, 3.14139, would still be of poor precision.

Fortunately, there are better alternatives. The German mathematician and designer of early mechanical calculators, Gottfried W. Leibnitz, had gone public in 1673 with the definition of π

using much faster converging infinite series of numbers, namely: $\frac{\pi}{4} = 1 - \frac{1}{3} + \frac{1}{5} - \frac{1}{7} + \frac{1}{9} \ldots$. Carried on long enough, this progression allows for any desired degree of precision, though more recent formulations with more rapidly converging terms are currently being used. At the start of the twentieth century, some 500 decimal places of π were known. Computers pushed that limit upwards, to the order of magnitude of 500 billion digits, and counting. Calculating π has become the litmus test for comparing the performance of new computer models with that of their predecessors.

Infinite series have been developed for most other irrational numbers, such as logarithms (log and ln) and trigonometric functions (sine, cosine, tangent, etc.).

Breaking the Rules

Squaring the circle is only the first of the three "unsolvable" problems of classical mathematics. Next in line is "trisecting an angle," the splitting of a given angle into three identical parts. Unlike the squaring of a circle, which is impossible for any size circle, tripartition can be done with certain selected angles, namely, the right angle of 90°, which we divide into three parts by subtracting a 60° angle, whose chord is known as equal to the radius. Likewise, we can split the angle of 27 (= 3 × 9) degrees into three 9° angles, using the 72 (= 2 × 2 × 2 × 9) degrees between a pair of adjacent radii of a regular pentagon, a figure of known design procedure. But that's about as far as we can get with ruler and compass alone. However, according to instructions from the Arabic *Book of Lemmas,* attributed to Archimedes (Fig. 1.6), adding a pin to our admissible drawing tools allows for trisecting any angle. First, we draw a semicircle around the vertex of the angle we wish to tripartite and mark the length of its radius R somewhere on the left of the ruler. With the pin placed at the intersection between the semicircle and the upper leg of the angle, we slip the ruler along the pin until the outermost mark coincides with the baseline and the other one with the semicircle. In that particular position, the angle between ruler and baseline is the desired one-third of the original angle.

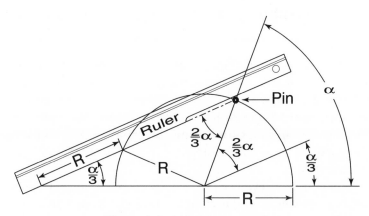

Figure 1.6   Trisecting an angle

This method can be understood from the markings in Figure 1.6, keeping in mind that both of the triangles left in the figure are isosceles. Those willing to build a working model of the device will have to compensate for the diameter of the pin by milling a $D/2$ deep recess into the right half of the ruler, so that the virtual extension of the ruler's straight edge on the left intersects the centerline of the pin.

The third and last of the classical "unsolvable problems" is the "Delian problem," which consists of finding the size of a cube relative to another of twice its volume (Fig. 1.7). The volume of a cube of side a is $V = a^3$, which makes $a = \sqrt[3]{V}$. A cube with double that volume, $V = 2a^3$, has the side-length of $\sqrt[3]{2a^3} = a \times \sqrt[3]{2}$, so that the problem boils down to duplicating graphically the irrational number $\sqrt[3]{2}$. Eratosthenes, best known for his measurement of the circumference of the earth (which we shall describe in Chapter 2), solved the Delian problem by means of the "mesolabium," a dedicated instrument he designed after all attempts at solutions within the framework of Euclidean geometry had failed.

We can build Eratosthenes' device by cutting from cardboard or stiff paper three right triangles with legs equal to twice the length of the cube's side. Next, we place those triangles (as in Fig. 1.8) on a straight baseline, place a ruler on top, and shuffle the

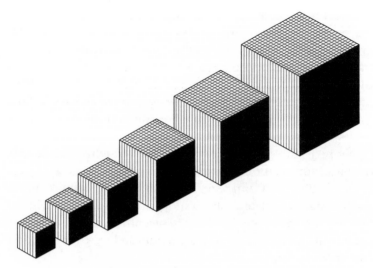

Figure 1.7 The Delian problem

triangles until the ruler's edge cuts all intersections as well as the midpoint of the vertical leg of the triangle on the far left. The upper segment of the leg of the center triangle then equals $a \times \sqrt[3]{2}$, which is the length of the side of a cube with twice the volume of one with side a.

Those yearning for mathematical proof of this little trick can derive (from Fig. 1.8) an expression of the type $b^3 = x \times y^2$, which for $x = 1$ and $y = \sqrt{2}$ becomes $b = \sqrt[3]{2}$, *quod erat demonstrandum.*

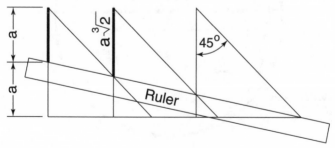

Figure 1.8 Erathostenes' graphic solution to the problem of doubling the volume of a given cube

Apart from these kinds of geometrical extravagancies, Eratosthenes, the leading scientist of his era, has been credited with the invention of the Hellenistic armillary sphere. Later developed into the astrolabe, this device is essentially a prototype of the geocentric cosmos (Fig. 1.9). Adjustable graduated brass rings replicate the great circles of the classical heavens: celestial equator, ecliptic (plane of Earth's orbit), meridian, horizon, and colures (the great circles linking the celestial poles and the points of the solstices and equinoxes). Astrolabes equipped with aiming sights became standard instruments for measuring the positions of celestial objects. Even Chaucer, of *Canterbury Tales* fame, wrote a book on the gadget. Made obsolete as observational instruments by the invention of the telescope in the seventeenth century, astrolabes became the precursors of the planetarium and were long used for educational purposes.

Without being part of the traditional unsolvable quests in mathematics, the problem of "fair division" deserves mention within this context. Half of a circle is 180°. But if two people are

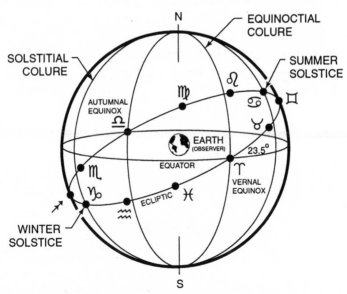

Figure 1.9 The great circles on the celestial sphere

cutting a cake, how can you make sure that it is cut exactly in half? For fair division, one person cuts the cake and the other gets to select which of the two pieces she or he wants to eat. That way, there's no reward to the cake slicer for doing anything else than cutting the cake in half. This has been proposed as a way to settle divorces: one partner divides the property owned by the couple into two lists; the other partner gets to choose which list to keep. The main math question then becomes: What happens when a *group* of people inherits something from a wealthy relative? What strategy could ensure fair division? Mathematicians are working on the answer.

Arithmetic and Geometry

It is possible that arithmetic and geometry have, as their common root, the surveying techniques of ancient Egypt. If so, then it is plausible that mathematical procedures can be graphically duplicated with the basic surveyor's tools: a pivoting string (equivalent to the modern compass) and a tightly stretched string (equivalent to the ruler) of determined length. In fact, the English expression "straight line" stems from the Old English words for "stretched linen."

Graphical solutions were held in high esteem in antiquity and endured as the science of nomography until their inherently limited precision made them obsolete with the emergence of digital calculators. Even so, professional exams for engineers, at least in the United States, still require knowledge of graphical methods.

In Euclidean geometry, that is, geometry on a flat surface, ruler-and-compass constructions have lengths or angles that are relative to the length of an initially defined straight line or angle. In that sense, they are unit independent. Units, displayed as the scales of our rulers, function as the link between geometry and arithmetic. For instance, Figure 1.10a shows how to slide a graduated ruler along a second one to simulate the geometrical equivalent of the arithmetical operation of addition: When the zero mark of one ruler is set on the first summand on the other, the sum can be read at the mark of the second summand. Shown here is the addition $5 + 8 = 13$. Figure 1.10b exemplifies the inverse process, the subtraction $153 - 116 = 37$.

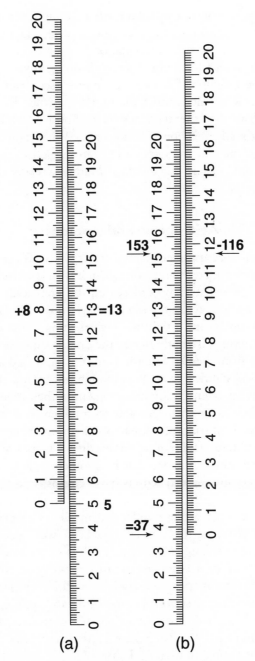

Figure 1.10 Graphic addition and subtraction

Though this result may seem pointless and graphic addition and subtraction had little practical value, the principal was picked up in 1620 by the English mathematician Edmund Gunter. Instead of ordinary markers, he used logarithmic scales, which convert multiplications into additions, and divisions into subtractions, according to the rules:

$$\log (a \times b) = \log a + \log b \quad \text{and} \quad \log \frac{a}{b} = \log a - \log b.$$

Likewise, scales calibrated to log sin α and log tan α can handle trigonometric functions.

Based on these principles, Amédée Mannheim, a French army officer, designed in 1850 the slide rule, which consisted of stock, slide, and cursor. Soon after its invention, it became the engineer's indispensable calculating tool until electronic calculators and computers took its place in the second half of the twentieth century. Unlike digital calculators, the slide rule is an analog device whose precision is limited by its length. Each additional decimal of readout precision requires a tenfold increase in scale length. Therefore, little could be gained by doubling or tripling the handy 10-inch length of the classical engineering slide rules, such as those made by Keuffel and Esser, the Rolex of the slide-rule world.

As a final example of the convergence of mathematics and geometry, we have graphical division (Fig. 1.11): the fractions $\frac{7}{12}$ and $\frac{5}{6}$ (equal to $\frac{10}{12}$) of a given length a can be found by drawing a straight line from the twelve-mark on an arbitrary yet regular scale to the endpoint of a, and then slating parallels to it from the marks for 7 and 10.

Ten-Based Units of Measure . . . a Rarity in Former Times

Today, we live in a base-10 environment, but as we have seen, the binary system of numbers has survived in many units of measure, especially in the 16 ounces of our pound. Sixteen, the fourth power of two ($2 \times 2 \times 2 \times 2 = 16$), is expressed as 10000 in binary. Likewise, the division of the inch into $\frac{1}{8}, \frac{3}{16}, \frac{1}{4}, \frac{3}{8}, \frac{1}{2}$, etc. is a purely

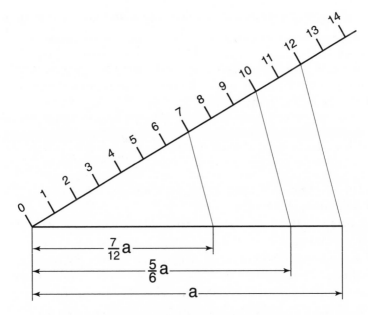

Figure 1.11   Graphic division

binary operation, as the denominators in all those fractions are powers of two. This helps in halving any given length; but adding or subtracting such figures invariably leads to the dreaded search for a common denominator.

For that reason, the once ubiquitous binary fractions in mechanical engineering have been replaced by thousandths. Instead of adding fractions of an inch such as $\frac{9}{16} + \frac{67}{128} = 1\frac{11}{128}$, engineers now sum up $0.563 + 0.523 = 1.086$. The result remains the same ($0.086 = \frac{11}{128}$), but the effort in getting it is reduced. This mix-and-match system (the use of thousands—a power of ten—with the inch) can be avoided by using the metric system, which uses factors of ten throughout.

· 2 ·

Going to Great Lengths

In ancient Egypt, where the yearly flooding of the Nile turned an otherwise arid region into first-class farmland, the law of supply and demand encouraged great intellectual achievements some five thousand years ago. As the retreating floods carried off some of the landmarks, there was a need to survey and redistribute the land among its owners. The Egyptians developed methods by which to assign areas to the landowners, based on data from the previous year. These techniques relied upon the fundamentals of geometry and arithmetics. Egyptians used a 10-based number system, although it lacked positional notation: the feature that in, say, the number 125 makes the one into 1×100, the two into 2×10, and the five into 5×1.

Ancient Units of Length

If surveying of their land holdings forced the Egyptians to discover the principles of geometry, it also necessitated an easily reproducible unit of length, the cubit. This unit was defined as the distance from a man's elbow to the tip of his middle finger. It wasn't only the Egyptians who used the cubit. From the Bible, in Genesis 6:15, we know that Noah's ark was to be 300 cubits long, 50 cubits wide, and 30 cubits high, or roughly 450 by 75 by 45 feet—not much space in which to store two animals of every kind for forty days and nights!

The cubit also turns up in the definition of the rabbinical mile, the distance one could walk on a Sabbath day without breaking the Talmud's injunction not to leave your home. In his *Rabbinical*

Mathematics and Astronomy, W. M. Feldman defines the rabbinical mile as 2000 cubits, which is approximately seven-tenths of a mile. As our foot is divided into 12 inches, the cubit was divided into twenty-four fingers. Four fingers made for a palm, three palms were a span, and two spans a cubit. Thus, a one-cubit-long string could be easily halved by folding it once to get a span, then into three equal parts for the palm, and finally twice into half, for the finger. Here again, the duodecimal system of numbers gets into the picture due to the divisibility of 12 by 2, 3, 4, and 6.

Setting the Standard

An early length standard in the shape of a $10\frac{1}{2}$-inch-long cubit rule, sculptured in stone, is part of the statue of the Sumerian king, Gudea, ruler of the city of Lagash from 2197 to 2178 B.C. Historians describe him as a "short, plump, jolly old elf, who built many temples. In one of his inscriptions, Gudea claims that under his rule, "the maidservant was the equal of her mistress, the slave walked beside his master, and the weak rested by the side of the strong." But some historians see that as an expression of wishful thinking, rather than a reality of the times.

In any case, Gudea's "cast in stone" length standard did not prevent different cubits from developing over time. In Egypt alone, there was the *common cubit* of six palms and the *royal cubit* of seven palms (20.6 modern inches), which was mostly used in architecture. The cubit was also adopted by the Hebrews, Babylonians, Greeks, and Romans, and although every nation's cubit was somewhat different from the next, they all measured between 18 and 23 of our inches. In ancient Greece, the cubit coexisted with the foot, which equaled two-thirds of a cubit, or 12.16 modern inches. Their foot was subdivided into four palms or sixteen digits. Six hundred feet made up one stadion.

Roman standards, the forebears of our customary foot/ pound/second system, were surprisingly accurate. Just as our present day foot comprises 12 inches, the Roman foot, measuring 11.68 modern inches, divided into twelve *uniciae* (parts). Five feet made for one *passus* (*pace,* double step), and the *mille passus* (mile) was 1000 paces or 5000 Roman feet long. As in Greece, a foot and a half equaled one cubit.

Even in modern times, the cubit has not been fully forgotten. India, while part of the British Empire, had a cubit (covid) of 18 inches, and so did Arabia. Data compiled by the United States for the 1960s mention the use of cubits in Burma, Somalia, and Macao.

Through the Middle Ages to the Renaissance

Around 300 B.C., Aristarchus of Samos came up with the idea that the sun was the center of our planetary system, yet the concept would not surface again until 1543, when a Polish monk, Nicolaus Copernicus, described a heliocentric system in his posthumously published book, *On the Revolutions of Celestial Spheres.* His claim that the Earth and the planets revolved around the sun, rather than the other way round, contradicted Ptolemaic teachings. More dangerously, Protestants and Catholics felt, it conflicted with Scripture.

Galileo Galilei, one of the early and most famous defenders of Copernicus's theories, was among other things a telescope designer and the discoverer of the four predominant moons of the planet Jupiter. In 1616 he received a formal warning from the Vatican that Copernicus's heliocentric theory did not agree with Scripture. This he ignored. Eventually he was summoned to appear before of the Holy Inquisition in Rome. He was spared the stake in exchange for summarily refuting under oath everything about the heliocentric theory, and he spent the rest of his life under house arrest for having "held and taught the Copernican doctrine."

Even though science and research progressed, technology remained somewhat rudimentary, and a pressing need for objective standard units did not arise before the onset of industrialization. Systems of measurements were still based on human body parts: feet, cubits, and the like, though they lacked precision and were subject to regional fluctuations. The idea of crafted standard bars must have caught on slowly. Unfortunately, such bars could be destroyed by war or natural disaster or, still more likely, falsified to favor regional interests or certain traders' cash flow. Precarious as body-parts-based standards were, they could be verified when needed. The length of a German foot, for example, could be reestablished any Sunday after church, should some local trader's yardstick seem suspiciously short.

Crafted standard bars gained ground only in the wake of Great Britain's worldwide trade expansion. Thus, in 1838 the English released the framework of the British Imperial System, which, rather than introducing new units of weights and measure, refined the existing ones. It still closely resembles today's American measuring system.

In continental Europe, split up into a variety of small kingdoms and dukedoms, the development of universal standards was delayed. Few states were willing to let go of their accustomed weights and measures for the sake of those from some other, possibly even hostile, country. In contrast to the British Empire, change on the continent was brought about by something more important than local interests, requiring a set of standards based on nature.

Thomas Jefferson's Idea

Ideas for a nature-based standard occurred long before the meter came into existence. Thomas Jefferson (1743–1826), credited for the invention of a hemp-treating machine, shooting stick, pedometer, and swivel chair, who would later become the third president of the United States, sponsored the size of the one-second solid-bar pendulum (Fig. 2.1), 117.357 modern inches, as a unified length standard. This is the length of a pendulum that takes one-second to swing from left to right, or vice versa. Note that the period of oscillation of a one-second pendulum is $T = 2$ sec.

The third president of the United States even planned a decimal system, subdividing twice the length of his one-second pendulum into ten new feet, of ten new inches each. But how long would a Jeffersonian foot be? If we use the symbols T for the period of the pendulum, L for its length, and g for gravitational acceleration, then the traditional formula connecting them is

$$T = 2\pi \times \sqrt{\frac{L}{3g}}, \qquad (2.1)$$

which gives, with $g = 9.80665$ m/sec^2, the length of the pendulum as

$$L = \frac{3g}{4\pi^2} \times T^2. \qquad (2.2)$$

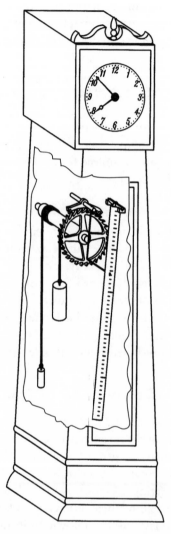

Figure 2.1 Weight-driven clockwork with solid bar pendulum and escapement

With *T* equal to the two seconds for the combined forward and backward swings of one second each, this length becomes 2.98086 meters. Multiply by 2 and divide by 10 to get for 10 Jefferson feet 5.96172 meters, or 23.471 inches per foot. Jefferson's inch thus measured 2.347 modern inches.

The basic shortcoming of the pendulum standard is that it requires counting seconds and fractions thereof. For a pendulum, any error in timing shows up doubly in its length because, according to the formulas shown above, an error of ΔT brings the pendulum's period to $T \pm \Delta T$ and the square (used in Eq. 2.2) to $(T \pm \Delta T)^2 = T^2 \pm \Delta T + \Delta T^2$. The last term becomes small enough to be neglected, so that we get $\Delta L/L \approx 2 \, \Delta T/T$. Because the percentage error in length is twice the percentage of the error in measuring time, the pendulum's period would have to be measured with twice the precision required for the length standard. To make things worse, eighteenth-century clocks were not particularly accurate to begin with.

Furthermore, a great number of other factors affect a pendulum's period of oscillation: gravity, barometric pressure, and ambient temperature, to name but a few. Even though Jefferson's pendulum standard didn't make it, it remains remarkable as a first step toward a unit based on one of the constants of physics rather than on dimensions of the human body. It is a curious coincidence that Jefferson was at the time the United States ambassador to France, and it would eventually be the French who succeeded in establishing a nature-based standard of length, the meter.

The Down-to-Earth Standard

Although the idea of standardization based on the size of the Earth gained acceptance in the seventeenth century, attempts to measure the Earth's circumference go back to ancient Greece. As early as the third century B.C., Eratosthenes of Kyrene calculated the Earth's dimensions by relating the north-south distance of the cities of Alexandria and Syene to their difference in latitude. The historic town of Syene (today's Aswan, known for its hydroelectric dam) is located close to the Tropic of Cancer, which explains why, on the day of the summer solstice, the Sun is almost directly overhead. According to tradition, Eratosthenes observed that the sunlight went straight down the shaft of a well. He realized that, at the same time, the Sun would not be overhead at, say, the city of Alexandria, almost straight north of Syene, because the Earth's surface is curved. The sunlight at Alexandria would strike a well, or a building, at an angle α, the difference in latitude of the two cities.

He derived that angle by looking at the length of the shadow cast by an Alexandrine obelisk and got a value of only 7°12′ or 7.2°. He figured if there are 360° in a circle, then the Earth's circumference is 360/7.2 multiplied by the distance between those two cities. In those days, Syene and Alexandria were separated by fifty days of travel by camel. The average camel traveled about one-hundred stadia per day, which amounts to 50 × 100 = 5000 stadia. So, he reasoned, the Earth must be 5000 × 360/7.2 = 250,000 stadia in circumference.

Eratosthenes ignored or did not know that Alexandria lies 2.5° west of Syene, which makes the north-south distance only $7.75/\sqrt{7.75^2 + 2.5^2} = 0.95$ of the distance as the crow flies.

Yet luck favored the brave, and the inherent incertitude in the distance measurements based on a camel's stride compensated for the neglected longitude detail. Comparison of Eratosthenes' figures with modern ones indicates an extraordinary precision of better than 1%. Even discounting such coincidental failure compensations, the accuracy of his results is impressive and would not be surpassed for close to two thousand years.

Sadly, as with many of antiquity's treasures of science, Eratosthenes' data on our home planet's dimensions didn't make it through the Middle Ages. Had Christopher Columbus been aware of them, he would never have believed he could reach the Indies by sailing westward from Portugal for a mere seventy days.

Attempts at a length standard based on the size of the globe could have started with Eratosthenes' results. Unfortunately, this idea did not surface again until 1670, when Gabriel Mouton, then parish priest of St. Paul's Church in Lyon, proposed the length of one minute of arc on a great circle of the Earth (meridian) as a new standard of length.

In 1735, an expedition was sent to measure the precise length of one degree of latitude as close to the equator as possible, and the following year, another expedition left for Lapland with the same aim. Their measurements must have given the French a pretty good idea of the size of the Earth, as they were meant originally for checking the oblateness of the globe by comparison with similar measurements at the observatory in Paris. One of the objectives was to check whether the globe deformed in accordance with Newton's laws of gravitation and (indirectly) with inertia.

A previous expedition, in 1743, to what is now Ecuador led to the development of the "Toise de Perou." This was the precursor of the modern meter, as half a century later the "provisional meter" was defined as 3 feet and 11.44 lignes of the Peru standard. This provisional meter was used in the final surveys of the length of the meridian quadrant by Pierre François André Méchain and Jean Baptiste Joseph Delambre.

In 1790, Charles-Maurice Talleyrand, who later became French foreign minister and Napoleon's appointee for grand chamberlain, called for a study of a universal system of weights and measures to be carried at the French Academy of Sciences. They decided to base the unit of length on that of the quarter meridian, which was to be derived much the same way as Eratosthenes had done it millennia earlier. Instead of Alexandria and Syene, they chose Dunkirk and Barcelona. Both are located at sea level, on the meridian that runs through Paris as well (Fig. 2.2), and at that time served the role of the zero meridian, although it would eventually lose out to the one that passes through the Greenwich Observatory in England.

In 1792, two renowned French scientists, P. F. André Méchain and J. B. Joseph Delambre, were trusted by the Bureau des Longitudes to carry out the precision surveying. Delambre's reputation rested principally on his tables of the positions of the Sun, Jupiter, Saturn, Uranus, and the Jupiter satellites, which were of sufficient precision to show irregularities in the orbit of the outermost known planet, Uranus. Those strange motions eventually led the English and French astronomers John Couch Adams and Urbain Jean Josef Le Verrier to predict mathematically the existence of an undiscovered planet farther out in the solar system. In short order, the German astronomer Johann Gottfried Galle found this new planet, later called Neptune, close to the spot predicted by the theory.

The discovery of Neptune in 1846 is lauded as a spectacular case of successful interaction between theory and observation. In hindsight, the orbit predicted by Adams and Le Verrier and the true orbit of Neptune aren't particularly close, so it was pure coincidence that, at the time of Galle's observations, the planet stood in a region where the mathematical and the real orbits almost overlap.

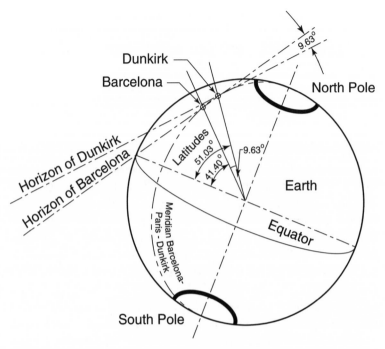

Figure 2.2   Surveillance of the meridian from Barcelona to Dunkirk

Delambre was entrusted with the measurement of the northern leg of the Paris meridian. Méchain, elected a member of l'Académie des sciences in 1782, got the southern part. This included
the stretch that cut through Spain, which, as it turned out, was
haunted by misfortune. In the Pyrenees, Méchain's group time
and again came face to face with the henchmen of regional revo-
lutionary committees. With France on the verge of war with Spain,
it was often hard to convince people that the elaborate collection
of precision surveying instruments in Méchain's entourage were
for exclusively peaceful ends. When war finally broke out, Méchain
was still in Spain, where he became prisoner. But this was not
Méchain's greatest personal tragedy. Released in Italy, an ally of
Spain, he reviewed his figures and found that the geodesic triangles of his survey did not exactly "close." If computed directly, the
latitude of a given spot in the city of Barcelona differed in three
seconds of arc from that computed at the 2-kilometer-distant

station of Monjuich. Rather than returning to Paris on schedule, Méchain remained in Genoa, going through his figures over and over again. Finally, in a desperate search for ways to overcome the discrepancy, he asked to extend the surveys down to the Balearic Islands. The French authorities agreed, albeit reluctantly. During that second part of his mission, though, Méchain contracted yellow fever. Tragically, he died in Castillion de la Plana in Spain on September 20, 1804.

When the dust had settled, the final meter standard, equal to 3 feet and 11.296 lignes of the Toise de Perou unit, became the metre (French for meter, which comes from the Greek μετρον = measure), and was thought to be the ten-millionth part of the quarter meridian (the distance from the equator to the pole). Because of their historical significance, the Toise de Perou that had served André Méchain as a length reference in his surveys and Joseph Delambre's Toise du Nord are still safeguarded in the Paris Observatory.

The efforts of Monsieurs Mechain and Delambre culminated on June 22, 1799, in the casting of a pair of meter bars as universal length standards. Compared to the 10^{-7}th part of the quarter meridian, those bars were short by 0.2 millimeter, because the researchers miscalculated the degree of oblateness of the globe. Nevertheless, those bars were the standards until 1889, when a new meter bar was made of a 90% platinum/10% iridium alloy (code named "1874 Alloy"), especially selected for its stability against aging and its low thermal expansion, only half that of pure gold. Since then, this worldwide meter standard has been kept in the reasonably stable environment of the basement of the Bureau International de Poids et Mésures in Sèvres, near Paris. It marks the meter unit as 39.37008 American inches long. As the British rhyme goes, "A meter measures three foot three, it's longer than a yard, you see."

In 1795, the new system was legalized in France and in the French empire of those times, and was readily adopted by government agencies and scientists. However, the business community and most of the population continued with the customary units, which varied from place to place. In 1812, Emperor Napoleon Bonaparte suspended the provisions of 1795, and it was

not until January 1, 1840, that the metric system became compulsory in France, its country of origin.

The Driving Force:
Onset of the Industrial Revolution

Although the Napoleonic Wars temporarily united the various countries of continental Europe, "L'empereur's" short-lived Paneuropa hampered rather than spread the metric system. The true catalyst for a unified system of weights and measures was the avalanche of inventions and discoveries brought about by the industrial revolution. In the first half of the nineteenth century alone, these included the first transatlantic cruise by the steam-powered vessel *Sirius* in 1838; George Stephenson's steam locomotive, "The Rocket"; and the construction of a railway from Stockton to Darlington, England, in 1825. Such projects resulted in the swift expansion of the manufacturing industry. As it's easier, quicker, and cheaper to fit standard-sized pegs into standard-size holes, it was not long before Sir Joseph Whitworth, in 1840, came up with the first unified screw thread system.

The massive progress in engineering and technology was the backdrop to the signing, in 1875, of the Treaty of the Meter by some seventeen nations, the United States being one of them. Some Swiss cantons had adopted the metric system as early as 1801, and two years later, in 1803, Milan, Italy, followed suit. The Treaty of the Meter provided for crafting a number of new, refined metric standard bars for distribution among its signatories, including the United States, which received the meter and kilogram standards in 1889. On April 5, 1893, these metric prototypes were declared "fundamental standards of length and mass" by Thomas Corwin Mendenhall, the Superintendent of Weights and Measures. Since that date, the yard, pound, and other measures have been defined in terms of the metric system, which made the United States an officially metric nation. The foot and the pound were no longer defined by national standard bars and weights, but by the corresponding metric units.

In this context, the yard (3 feet) was redefined as 3600/3937 meter and the pound avoirdupois as 1/2.2046 kilogram. After

World War II, the yard/meter relation changed slightly to make the inch equal to exactly 25.4 millimeters, some 0.0000508 mm less than the original conversion. In the process, the foot also received an exact metric conversion factor of 0.3048 meter.

Back to a Nature-Based Meter

Ironically, while the metric system took hold, the meter's condition as a nature-based standard of length eroded. As we now know, the quarter meridian measures 10,001,954.5 meters rather than the straight ten million targeted by the metric system's founding fathers.

Actually, it is irrelevant whether or not the relation of the length of the meridian quadrant to the length of the meter bar can be expressed by an integer figure. What matters is the existence of an accepted, reproducible length standard, but unfortunately, even that had not been accomplished. Periodical checks on the venerable meter bar with instruments of successively higher accuracy revealed variations of a few ten-thousandths of a millimeter. This led the American National Institute of Standards to redefine the meter as 1,650,763.73 times the wavelength of the orange-red light emitted by the radioactive isotope krypton-86.

But the story does not end here. Ask an American how far a town is, and the answer might be "a half hour": you had asked for a distance but were given a time. Your friend knows roughly what the speed of driving is. Likewise, physicists now know the velocity of light in a vacuum is an invariable constant throughout the cosmos. This makes it the ideal length reference, and in 1983, the Conférence Générale des Poids et Mesures (CGPM) introduced the definition of the meter as exactly "the length of the path traveled by light in vacuum during a time interval of 1/299792458 of a second." (Of course, the denominator of that fraction is the velocity of light in meter per second). That is, if you want a length, you're given a time. In doing so, the meter had gone full circle: from a nature-based unit to a man-made bar, and back to a nature-based unit once more.

And no doubt old Jefferson would be proud that his concept of a time-related length standard had won out, after all.

· 3 ·

Degrees of Separation

Angles and Solid Angles

According to ancient legend, Helen—wife of Menelaus, king of Sparta—was one of the most beautiful women in history. Her beauty attracted a young man by the name of Paris, who abducted Helen and took her to his home city of Troy. Menelaus sought to bring his queen back and so launched a fleet of ships in hot pursuit of Paris. Thus, Helen came to be described as possessing a "face that launched a thousand ships." Homer's classic tale *The Iliad* tells the story of the ensuing Trojan War.

Troy, then a blooming commercial port on the northeastern edge of Asia Minor, controlled from about 1500 B.C. through 1100 B.C. the sea lanes that the bulk of Greek traders depended upon. But then, as now, the Aegean Sea that the Spartans had to cross was both treacherous and rock infested. On September 26, 2000, the lives of seventy-seven crew and passengers were lost on the Greek ferry *Express Samina,* which struck a rock and sank off the island of Paros. If such radar, GPS, and auto-pilot equipped ships still run astray, how did prehistoric boats, such as King Menelaus's fleet, make it through the approximately 450 statute miles to the tip of Asia Minor, powered by nothing more than the free, environmentally friendly, renewable energy of the wind?

To answer this question, we would normally have to rely on indirect evidence, drawn from the excavations of objects such as statuettes and tableware, hammered out of massive pieces of gold. But the poet Homer, who is thought to have lived and worked

some three hundred or four hundred years after the war, describes in *The Iliad* the Spartan fleet of wooden warships as sailing unerringly toward Troy. But in *The Odyssey,* the sequel to *The Iliad,* it takes the hero Odysseus some ten years to make the trip back home. While clearly a literary device—just as the Hebrew people are described as wandering in the desert for some forty years though one can cross that particular desert on foot in only a few days—it does show the perils and unpredictability of ancient seafaring.

Most ancient peoples sailed by use of the stars. Phoenicians set up trading routes throughout Europe, and peoples from Polynesia could sail vast distances across the oceans on their reed boats, as Thor Heyerdahl confirmed in his great voyage on the *Kon Tiki.* Navigational technology did not change much for many centuries. In the days of Columbus, seafarers relied a great deal on the proverbial "three L's"—log, lead, and lookout. The log—basically a piece of wood on a cord—was thrown overboard at the ship's bow and watched by the crew as it drifted toward the stern. From the time that elapsed and from the length of the vessel, sailors calculated how fast they were traveling. Combined with astronomical latitude measurements, they knew in what direction they were headed. So, knowing where they were headed, how long they had sailed on that heading, and how fast they were going, they had a good idea of roughly where they were. This method to "guesstimate" the ship's position was known as "dead reckoning," a method widely employed at times when chronometers for checking longitude were things of a distant future. However, sea currents tended to distort the results. Even today, the students at the United States Naval Academy are still taught the principles of dead reckoning. If your helicopter is shot down and you have no instruments to help you find home, dead reckoning is one of the best ways to get there. All you need is a watch.

The second "L," a simple piece of lead used for depth measurements, was developed into a pocket and filled with sticky tallow to let sailors bring up samples from the sea floor. From the contents of the pocket, such as white or colored sand, gravel, soft worms, or shells, a seasoned mariner could sometimes guess where he was.

It is hard to believe that such simple tools were not available in antiquity, even if history fails to document them. After all, such

civilizations produced Ptolemy, the ancient astronomer whose mathematics for predicting the positions of Sun, Moon, and the planets endured all the way through the Middle Ages until Kepler's laws took their place.

The compass was then still two millennia in the future. The earliest written reports on the use at sea of a rudimentary lodestone direction finder come from Chinese mariners in A.D. 1115. The Vikings are also thought to have used lodestone, a magnetic rock that was suspended from string to serve as a compass, which would have helped them sail to Greenland and America. The advantage is that the lodestone would always point north. Stars can only help you if the skies aren't cloudy, and in the North Atlantic—where 30-foot waves are not uncommon—great visibility is not guaranteed. It was not until the year 1300 that the wind-rose compass, mounted in a set of gimbal rings so that the instrument would remain horizontal regardless of whether the ship was tilting and rolling in heavy seas, was put into use.

Menelaus's sailors may have steered by the stars, but they didn't have it all that easy. Unlike scouts who simply have to find Polaris, the Spartans had two main problems. First, they imagined the earth as flat, so that the definition of north as the direction leading to the pole was meaningless. And second, the North Star, Polaris, then stood far from the celestial pole. The pole star back then would have been one of the lesser prominent stars in the constellation Draco the Dragon.

The problem with Polaris "moving" is not that the stars have traveled across the sky that much. The night sky appears pretty much the same tonight as it did three thousand years ago. It's just that the Earth's axis has moved. If you spin a top, it doesn't usually land with its axis pointing upward, but it lands at a slight angle. The axis then gently and elegantly traces out a cone, a motion technically described as "precession."

All of the planets in the solar system have a precession. The Earth's rotation of one turn per day makes it act like a real big gyroscope whose axis tilts at an angle of $23.5°$ to the normal of Earth's orbital plane, the ecliptic. Just as the swash of the axis of a top goes far slower than its rotation, the Earth's axis takes 25,800 years to swing full circle. That's why the vernal equinox, marked in antiquity by the Sun's pass through the constellation

of Aries the Ram has since moved into the next symbol of the Zodiac, the constellation Pisces the Fish.

Likewise, the celestial pole, the center of the apparent daily motion of the stars, precesses during the same timespan around a circle of 23.5° radius over the celestial background. At present, we are fortunate to have Polaris no farther than one degree of arc from the celestial pole, and, unlike most of the world's stock markets at present, things will get better before they get worse. By 2102, Polaris's distance from the pole will be down to less than half a degree, which is as close as this star will ever get. When ancient Egyptians were busily building their pyramids 4600 years ago, Alpha Draconis, the star at the tip of the snout of the heavenly dragon, was the pole star, while Vega (Alpha Lyrae) had this honor some 12,000 years ago. This bluish star held the record as the brightest star in the Northern Hemisphere through the early decades of the twentieth century, but since then it has given way to the reddish Arcturus. Neither of these stars is classified as variable, however. Increased absorption of light by the Earth's atmosphere, more pronounced in the blue than in the red, is probably the culprit responsible for Vega's demotion.

Even on a flat Earth, under a sky missing the polar star, ancient seafarers could have deduced north (without defining it) as the center of the circles described by circumpolar stars. On the other hand, south was marked by the meridian passage of the Sun, though the Sun's path—horizontal at noon—makes it hard to determine with reasonable precision the moment it reaches its highest point over the horizon, unless one has a watch. By contrast, east and west can be recognized by the points on the horizon where equatorial stars, such as the belt of Orion in the constellation of the hunter, rise and set.

Now, there are 180 degrees in a triangle, so the height of the celestial pole over the northern horizon is equal to the latitude of the observer's location. Conversely, the angle to the highest point of the celestial equator is 90° minus latitude. That's where the Sun stands at noon on the days of the equinoxes. For the city of New York, at about 40° latitude, the high point of the celestial equator arcs by 50° over the southern horizon. With 23.5° for the obliquity of the ecliptic, we can figure the stand of the Sun on the

longest day of the year as 50 + 23.5 = 73.5°, and as a puny 50 − 23.5 = 26.5° around Christmas.

Although the Sun comes to mind first when we think of finding latitude, one needs tables on its daily wandering around the Zodiac in order to do so. Seafarers in the fifteenth century used rudimentary instruments to "shoot the Sun." They resemble a clock face, divided into 360°, with one single diagonal hand that carries the viewing sight and crosshair on opposite ends. Some instruments were suspended from a hook in ways that their own weight leveled them into a horizontal position. In others, the pointer was kept in the vertical by a counterweight at one end, while the observer rotated the sight and crosshair-equipped graduated circle to shoot the target.

Luckily for Homer's heroes nearly a thousand years earlier, the people of Sumer and Babylonia (the regions of present-day Iraq) had divided the circle into 360°, which we still use today (Fig. 3.1). It stems from the sexagesimal number system of ancient Babylon. Each degree has sixty minutes (60'), and each minute has sixty seconds (60"). As 360 has so many divisors, the right angle (90°) can be halved or divided into two, three, five, six, nine, ten, or fifteen parts with integer numbers of degrees. This great utility is why the degree survived into the present and is even enshrined in the International System of Units. By international convention, angles are measured counterclockwise (CCW), but as

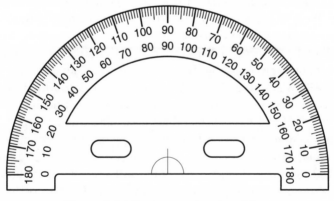

Figure 3.1 Protractor, showing degrees

Figure 3.1 shows, the protractor has both—clockwise and counterclockwise numeration—for its users' convenience. For the ancient Babylonians, who had a keen interest in astronomy, the 360° in a circle would be conveniently close enough to the 365 days in a year in order to take the daily advance of the Sun along the Zodiac as one degree.

There are many exciting stories of people traveling vast ocean expanses in open boats. For example, on April 28, 1947, Thor Heyerdahl and his five crew members embarked from Peru on a balsawood raft that was built according to the traditions of South America's pre-Colombian Indians. After 101 days on the open sea, they reached the Polynesian island of Raroia.

The ancients could have steered by the stars, but the great voyages of exploration in the seventeenth and eighteenth centuries were dominated by use of the sextant, which navigators used to fix the ship's position and to chart its course. Figure 3.2 shows such an instrument with a readout precision of 0.1°, that is, six minutes of arc. Exact readouts of typically one-tenth of a division are possible by reading the results from a sliding scale, the "Vernier" (see detail on Fig. 3.2), which is spaced to divide nine degrees on the main scale into ten parts. Start by reading the instrument's scale at the zero mark of the Vernier; for tenths, count the spaces between the Vernier zero and the line on the Vernier that coincides with a notch of the true scale. Finer divisions, such as 0.02 millimeter with a Vernier that divides 49 millimeters into fifty parts, are possible, but numerical readouts have made this ingenious device obsolete now. We have thus moved one more thinking process from the brains of the beholder onto a chip.

With the sexagesimal division of the circle, the related arithmetic tends to become confusing. If you're charting a course toward latitude 43°13'25" from 22°54'39", you must operate with a carryover of sixty rather than from the customary ten, in order to get the bearing of 20°18'46" that you must steer. The alternative of converting minutes and seconds to degrees by division through sixty and 3600, respectively, cuts down this entire operation to 43.22361 − 22.91083 = 20.31278°. This result can be retrofitted, multiplying the decimal fraction of 20.31278, that is, 0.31278, by 60 to get 18.7668 minutes of arc, and repeating that process with the fractional part of minutes, which gives 0.7668 × 60 = 46", as above.

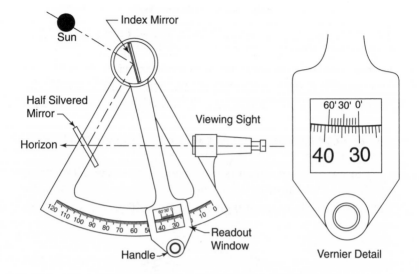

Figure 3.2 Sextant with Vernier readout

Nowadays, though, ships use the global positioning satellite system and are controlled by computers.

On the other hand, arc minutes and seconds are still popular as stand-alone entities, for instance in the definition of the astronomical unit of parsec as the distance from which the apparent radius of the Earth's orbit (1 AU) subtends exactly one second of arc. Astronomers also use seconds of arc to express the spacing of binary stars, and minutes of arc to describe the apparent dimensions of nebulae and galaxies. The apparent size of the M51 galaxy in Canes Venatici (the hunting dogs), for instance, is cited in Burnham's *Celestial Handbook* as 10.0' × 5.5'. Amateur astronomers often rely on the rule of thumb that one centimeter (approximately 3/8 inch) at arm's length covers one degree of arc on the celestial sphere. Before the introduction of the meter, Gabriel Mouton, the parish priest of St. Paul's Church in Lyon, proposed one minute of arc of the meridian as a new unit of length. Its acceptance lives on in the definition of the international nautical mile as one minute of longitude at the equator. One nautical mile per hour, as every sailor knows, is a speed of one knot.

Degrees and Radians

The circumference C of a circle is $2\pi R$, where R is the radius. That is to say, there are 2π radii in the circumference. By analogy, the famous 360° in a circle can be thought of as being 2π radians. If 2π rads = 360° then one radian is $360/2\pi$ = 57.296°. While the degree figures as an alternative unit in the International System, the radian (Figs. 3.3 and 3.4) is the basic SI unit of a plane angle. A right angle is equivalent $\pi/2$ rad, a 45° angle converts to $\pi/4$, and an angle of 30° to $\pi/6$ (compare Fig. 3.3).

There have been attempts to metricate the degree. The grad, seen in the outermost circle of Figure 3.3, is one-hundredth of a right angle, so there are 400 grads in a circle. They can sometimes

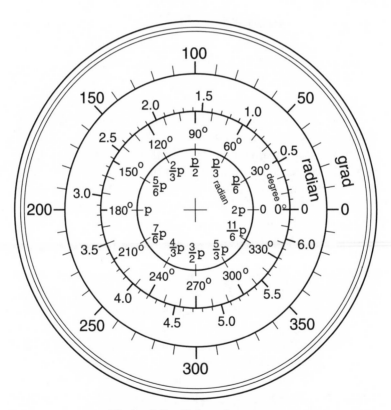

Figure 3.3   Universal protractor

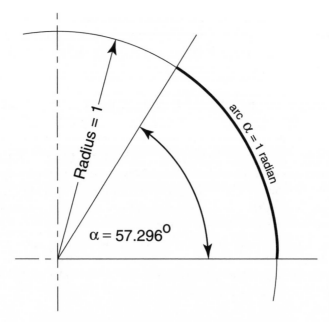

Figure 3.4 Equivalence of degree and radian

be found on pocket calculators and in trig tables along with degree and radian. Although the grad never caught on in the United States, it was legalized in 1937 in Central Europe as a surveyors' unit. Since then, its name has changed to the gon, defined within the guidelines of the International System as $\pi/200$ radian. Grads are also used in maritime navigation to calculate the course and define the position of a vessel at sea. Mind you, there is a problem: charting a course of 30° is much simpler than steering on a bearing of 33.33333 gon.

Where the Triangle Fits In

While a protractor (Fig. 3.1) allows one to draw and measure angles on paper, surveyors employ an instrument called the theodolite for their purposes. The first mention of theodolitus comes in a work by an Englishman, a Mr. Digges, in 1571, though the instrument didn't catch on in Europe until the eighteenth

century. The editors of the Oxford English Dictionary don't seem to know how the word was constructed. Be that as it may, the theodolite is a crosshair-viewing telescope, mounted to swivel around the centers of a horizontal and a vertical graduated circle, respectively.

In the absence of such instrumentation, an angle can be defined with the help of the one and only polygon whose shape is unequivocally determined by the length of its sides: the triangle. For any other nonregular polygon, the length of the sides and one or more angles must be given: four sides and one angle for a quadrangle, five sides and two angles for a (nonregular) pentagon, and so on. Therefore, the triangle has become the basic geometrical figure in surveying, as well as the load-carrying element in geodesic domes. A right angle, for instance, can be laid out by use of a triangle, with the sides three, four, and five units of length, for it complies with the Pythagorean theorem for right-angle triangles, in this case $3^2 + 4^2 = 5^2$.

Trigonometric Functions of Angles

In principle, triangles of any kind could be used to define angles, but equilateral triangles were the norm before the advent of trigonometry. Scaling the triangle in Figure 3.5 by $1{:}\ell$ makes the length of its legs equal to $\ell = 1$. In such an "isosceles unit triangle," the length of chord $\alpha = s/\ell$ unequivocally defines the angle between the two equal sides. The corresponding value in degrees can be looked up in mathematical tables, which were compiled with help of the infinite progression

$$s/\ell = 2 \times \left(\alpha - \frac{\alpha^3}{3!} + \frac{\alpha^5}{5!} - \frac{\alpha^7}{7!} + \frac{\alpha^9}{9!} - \cdots \right).$$

The factorial numbers in the denominators of the terms of this series make for its fast convergence.

As another option, we may define a by the related tangent on the unit circle, as tangent $\alpha = s/h$, also shown in Figure 3.5. Tables for chord and tangent in relation to the subtended angle α reach back to Ptolemy's *Almagest* from the second century A.D., which

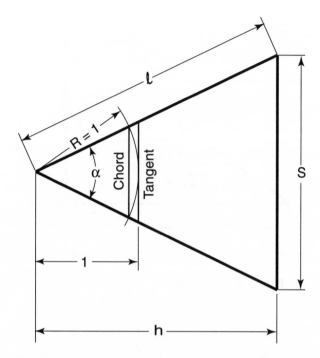

Figure 3.5 Defining angles through equilateral triangles

lists them to five decimal places of accuracy for angles in 30 minutes of arc intervals.

Trigonometric Tables

Present-day trigonometric tables still resemble those of the *Almagest*, except that instead of listing chord and tangent in the Ptolemaic sense, they halve the angle α and consequently the numerical values for chord and tangent, now tagged with the labels sin and tan. Figure 3.6 shows them as well as the arc function. You can see that

$$\text{arc } \alpha = \frac{a}{\ell} \qquad (3.1)$$

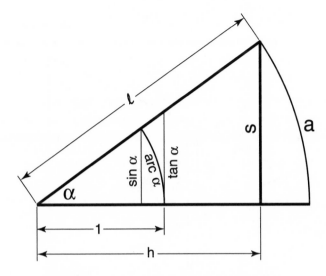

Figure 3.6 Trigonometric functions of angles

$$\sin \alpha = \frac{s}{\ell} \qquad\qquad (3.2)$$

$$\cos \alpha = \frac{h}{\ell} \qquad\qquad (3.3)$$

$$\tan \alpha = \frac{s}{h} \qquad\qquad (3.4)$$

These so-called trigonometric functions are basic in mathematics and calculus.

A typical application is shown in Figure 3.7. A surveyor sets up a theodolite at a known distance d from the base of the object (the Eiffel Tower). He then measures the angle α to the top of the tower. Knowledge of d and α allows him to calculate the height h, which is $d \times \tan \alpha$.

Another problem along the same lines—yet solved with the sin function—is to determine the slope of a piece of land. Perhaps there's a problem with the drainage of a property, or the likelihood of landslides or avalanches on a steep mountainside; or maybe you just want to know how steep your driveway really is. A surveyor would put a marked pole into the ground, then measure along the

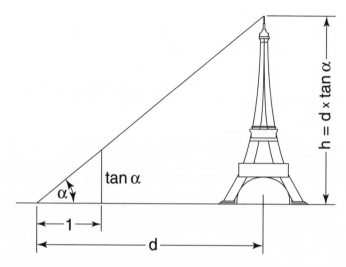

Figure 3.7 How to find the height of a tower (without climbing to the top)

surface of the hill, a distance ℓ away. She can measure $H_1 - H_2$, the drop in height over distance ℓ. The equipment-challenged surveyor can use a clear garden hose with water in it (Fig. 3.8). Whatever the choice, the slope is given by $\sin \alpha = (H_1 - H_2)/\ell$.

The Kamal, an ancient Arabian navigational tool consisting of a carved mahogany transom with a cord attached at its center, reads angles by measuring their respective tangents. You can make one by drilling a hole through the center of an (expired) credit card, and pull a length of thread through it that has a knot on one end. With a $3\frac{3}{8}$ inch (3.375 inch) wide card, we can use this "instrument" for a variety of tasks, such as finding the date when the moon stands lowest at its meridian transit. For the city of New York, the lowest full moon culminates in summertime about $22°$ above the horizon. A calculator gives us the tangent 0.194 of half this angle ($11°$). With $s = 3.375/2 = 1.6875$ inches and $\tan \alpha = s/h$, reshuffled into $h = s/\tan \alpha$, the respective thread will be $h = 1.6875/0.194 = 8.68$ inches long. Secured at that distance, our credit card should just cover the angle between moon and horizon. The instrument should be used by keeping the cord lightly between your teeth

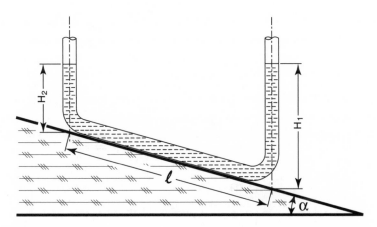

Figure 3.8 A clear garden hose doubles as a theodolite in finding the inclination of a steep driveway.

while pulling the transom forward until it fits the angle to be checked. For best results, make sure the card is at a right angle to your line of sight.

For wider angles, such as the height of Polaris (equal to the latitude of your point of observation), you will nedd a bigger transom to allow for a thread of reasonable length.

Dimensions of Trigonometric Functions

As we have seen, trigonometric functions, such as $\sin \alpha = s/\ell$, $\cos \alpha = h/\ell$, $\tan \alpha = s/h$, are defined by variables that come in units of length (meter or foot). That makes their ratio, length/length, dimensionless.

Spatial Angles

There are 2π radians in a circle, but let's leave the confines of the plane. Imagine a sphere of radius 1. Now, there are 2π steradians in a hemisphere. (Because a picture is worth a thousand words, see Figure 3.9.) The International System of Weights and Measures (SI) defines the steradian (sr) as the ratio of the area to the square of the distance:

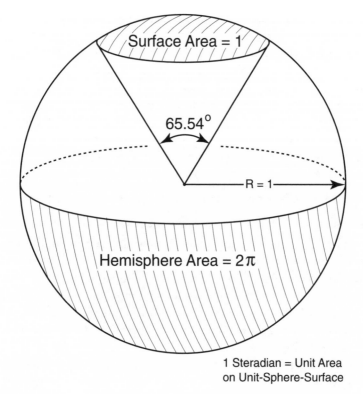

Figure 3.9 Spatial angles: the steradian as surface area of the unit sphere zone

$$\alpha = \frac{A}{R^2}. \qquad (3.5)$$

Accordingly, the surface of a sphere with radius = 1 is 4π steradians, and one steradian covers $1/4\pi = 1/12.566 = 0.0796$ of a unit sphere's total surface area of 4π. For a circular spherical segment, one steradian subtends the angle of $65.54°$.

The Inverse Square Law

The surface area S of a sphere of radius R, $4R^2\pi$, relates to the surface area of the unit sphere S_0 (with $R = 1$) as $S/S_0 = 4R^2\pi/4\pi = R^2$. That applies to heat, light, radio waves, etc. If you sit twice

as far from a radiator as you did earlier, you get heat at only a quarter of the previous rate.

If Sirius, the brightest "fixed" star in the sky, which is 8.5 light-years distant in the constellation of Canis Major (big dog), were moved ten times farther out, then $10^2 = 100$ times less light would reach us from the star. Its luminosity would thus decrease by five orders of magnitude in the conventional scale of star brightnesses. This would take it from -1.44 to 3.56, so it would drop from first to well below 293rd in a ranking of brightest stars.

Horse Sense and the Size-to-Volume Ratio

The term "cube" for the third power stems from the $V = a^3$ relation between the volume V of a cube and the length a of its sides, in the same way as the mathematical term "square" is derived from the expression $A = a^2$ for the area A of a square. The term "zenzizenzizenzic," for the eighth power of a number, was introduced by Robert Recorde, a Welsh mathematician of the sixteenth century. It never caught on. Recorde also introduced the symbols =, >, and < in his book *The Whetstone of Witte*, published in 1557. These did catch on. He had other interests too—he was a medical doctor and wrote the book *The Urine of Physick*, a sixteenth-century manual on urinalysis.

Because volumes go as R^3 and areas as R^2, we can draw some interesting conclusions. Eohippus, the ancestor of the modern horse, measured a sparse half meter (about 20 inches) in length as it roamed the Earth some 40 million years ago. Toward the Miocene, 24 to 5 million years ago, a small pony, the Merychippus, lived and survived into the early Pleistocene, 5 million years ago. Then came the larger Pliohippus and finally Equus, which includes the horses we know today.

Even though each protohorse seemed to get successively larger, did they really evolve smoothly from one into the other? Professor Othniel C. Marsh of Yale thought so, and in 1874 this belief led to a famous exhibit of horse skeletons in the American Museum of Natural History, each larger than the previous one and with further developed hoofs.

Marsh's views of this evolution endured for almost a century.

But in 1951, George Gaylord Simpson's book *Horses* told a different story; according to Simpson, the rates of development had not been gradual but jerky. Sixteen different genera had actually developed, of which fifteen became extinct.

Mathematics bears Simpson out. For instance, geometrical doubling of the original Eohippus's size would bring its volume, and thus also its weight, to $2^3 = 8$ times of what it was originally, but the cross-sectional area of the animal's bones would increase by a mere $2^2 = 4$ times. Compressive stress (weight/area) on the greater horse's bones would then become $2^3/2^2 = 2$, double the former value. Thus, the "new horse" would need bigger and maybe stronger bones. It would also be expected to have roughly $2^3 = 8$ times the amount of blood of its predecessor, and pump it twice as high. This would demand a heart of $2 \times 2^3 = 2^4 = 16$ times greater pumping capacity, or power, housed in a body only twice the size of its predecessor.

Thus, the mutual proportions of all its body parts would have to undergo fundamental changes in relation to body size. If not, premature extinction would await.

Surface-to-Volume Ratio

A cube's surface area is $S = 6a^2$, so the ratio of surface to volume is

$$\frac{S}{V} = \frac{6a^2}{a^3} = \frac{6}{a}.$$

Again, this leads to interesting insights. The food-storing capacity and the energy available to a warm-blooded being increases with its body volume, but the heat loss into the environment by convection and radiation depends on the external surface area of its body. Consequently, smaller animals lose more heat through their relatively greater skin surface than bigger ones. To overcome this problem, they have greater appetites to fuel them, but it also sets a lower limit for the size of warm-blooded creatures.

The Pigmy Shrew, a mouselike mammal common in Utah and New Mexico, whose body, without the tail, is only about 2 inches long, must feed almost continuously day and night just to keep

warm. It consumes the equivalent of its own body weight every twenty-four-hour period. Left without food, the shrew will die within two to three hours.

On the other hand, the large size of the dinosaurs makes it likely that the energy produced by their high-volume bodies from food intake, combined with the low overall heat loss through the comparatively small area of their skin, kept their body temperature within the range of that of modern warm-blooded vertebrates, though they did not have the latter's built-in climate-control system.

· 4 ·

The "Obvious" Unit of Time

As most children can tell you, there are 60 seconds in a minute, 60 minutes in an hour, and 24 hours in a day. The second, an inherently nature-based unit of time, is therefore 1/86,400 the length of the average solar day. The division of the day into 24 hours looks like a carryover from the duodecimal system of numbers, while the 60-based hour/minute/second relation stems from the sexagesimal number system, popular with ancient Babylonians.

Hebrew calendar makers of antiquity used an entirely different approach. They divided the hour into 1080 parts, to create their basic unit of time, the *helakim*. With 3600 seconds to the hour, one helakim can be thought of as approximately 3 seconds.

The use of the number 1080 has an interesting history, for it comes from the duration of the precessional year. Just like a spinning top "wobbles" or precesses around its axis, so too does the Earth. This precession manifests itself in the westward motion of the equinoxes. In antiquity, the vernal equinox used to occur when the Sun was in the constellation of Aries the Ram. Now it is in the preceding constellation, Pisces the Fish. It takes 25,800 years to go around the Zodiac, and thus it will return to its ancient place in Aries in 23,400 years. The magnitude of precession, historically taken as one degree of arc per seventy-two years, or $72 \times 360 = 25,920$ years for a full cycle, led the Hebrews into using the ratio of that number, and the number of hours per day, $25,920/24 = 1080$, as the key for deriving the hour's subunit, the helakim.

Such units make up part of today's official Israeli calendar. For instance, the lunar constants for determining Tishri 1, the first day

of Rosh Hashanah, the Jewish New Year, are given for the first
lunation (New Moon) as four weeks, one day, twelve hours, 0793
helakim; and for the twelfth lunation as fifty weeks, four days,
eight hours, 0876 helakim, etc. Tishri 10 marks Yom Kippur, the
Day of Atonement, which is the holiest day of the Jewish year and
is dedicated to reflections on interrelations with God and with
one's fellow men, and to atonements for one's wrongdoings. One
of the most colorful Bible stories is linked to Purim, the Feast of
Lots, on the 14th Adar, in remembrance of the Babylonian exile.
Haman, adviser of King Ahashveiros, obtained an order to exter-
minate the Jewish population, yet his plans were spoiled by the
righteous Mordechai and Esther.

The plethora of different calendars used since the dawn of civilization stems from attempts of housing under one roof the three
natural measures of time: the day, the year, and the time between
New Moons. The Gregorian calendar solved the problem of fit-
ting the 365.2422 days into the year with the introduction of an
extra day (February 29) every four years, except for centennials
not divisible by 400. Thus, we got an extra day in 2000, but we
won't have that bonus in 2100, 2200, and 2300. A still more precise rule would somehow have to distribute 2422 leap days within
the next 10,000 years.

Fair enough, but how about the 29.5306 days of the Moon's
period, which makes for 12.368 months per year? The Sumerians
of Babylonia counted twelve lunar months as a year and inserted
an extra month in the calendar about every four years. This calendar was adopted by early Egyptians, Greeks, and Semitic peoples. Early Romans also based their calendar on the moon. In our
time, movable feasts (Easter and Passover, Whitsunday, Rosh
Hashanah, Israel's Independence Day, etc.) are reminders of the
incompatibility of the Earth's and the Moon's's periods of revolution, but otherwise the Moon's importance for people's survival
after sunset has dwindled, and so has its role in calendar making.

As with the partition of the circle into 360°, the unit of time
has resisted metrication and in all probability this will never be
changed. The only serious attempt at a decimal division of the day's
length occurred during the Enlightenment. Under the banner of
the French Revolution, the Académie des Sciences endorsed, in

1793, Pierre Sylvain Marechal's metricated calendar and a metric clock face (Fig. 4.1). Both remained French legal standards until 1806, when the Gregorian calendar was reestablished.

Marechal's calendar divided the year into twelve months (Vendémiaire, Brumaire, Frimaire, Nivôse, Pluviôse, Ventôse, Germinal, Floréal, Prairial, Messidor, Thermidor, and Fructidor) of thirty days each, with five days left over, or six days in leap years. The five "leftover" days were called "sansculottiden" to honor

Figure 4.1 A metric clockface

those who refused to wear the French *culottes* (breeches) traditionally worn by French aristocrats.

Each month of the "enlightened" calendar consisted of three decades of ten days each, and the day was supposed to have 10 hours of 100 minutes each. But the sans-culottes' haste in imposing their new system was not met with equal zeal by the new society's watchmakers, and a dearth of 10-based timepieces made the metric watch and calendar short-lived.

Between the two world wars, the League of Nations proposed a calendar of thirteen months of twenty-eight days each. Unlike the French calendar, this one left only one day ($365 - 13 \times 28 = 1$) unaccounted for, which was to be the first day of the year. Although the idea makes more sense than the calendar we actually use, it failed to catch on. Unswerving, the World Calendar Association and certain commissions within the United Nations and UNESCO still follow up on new calendar ideas, while the Vatican wishes to eliminate the varying dates of Easter Sunday among the Orthodox and Catholic churches.

When the dust finally settled, the second prevailed as the basic unit of time, although decimal fractions of minutes are occasionally employed in industrial time studies. Yet as the definition of the second involves a fraction of the average solar day, the latter should, at least nominally, rightly carry the "basic time unit" title.

Apparent and Average Solar Day

We are all familiar with seasonal changes in the length of daylight, and the dreaded days when the clocks are set forward or back, instituted as an energy-saving measure. But the total length of a day also varies in the course of the year, albeit to a much lesser degree. A day and a night can therefore add up to a tad more or less than 24 hours. Scientifically, the term "solar day" includes the sunless night and is defined as the time span between two successive culminations, or meridian transits of the Sun. At meridian transit, the Sun stands exactly south.

The day's length is not a consistent 24 hours, because the apparent motion of the Sun over the sky is the vector sum of two distinct

velocity components: one (with a period of 23 h 56 min 4.1 sec) stems from Earth's rotation; the other (with a period of 365 days 5 h 48 min and 45.5 sec) from Earth's orbit around the Sun.

If you were at the North Pole, both these motions would be counterclockwise. An analogy may be useful: imagine a car passing a bus stop at 40 miles per hour (mph). A cyclist, pedaling at 10 mph in the same direction, sees the car pulling away from him at only 40 − 10 = 30 mph. As with the car and bicycle, the daily and yearly motions of the Sun subtract from each other. The Sun's daily sweep over the sky lags behind that of the rest of the celestial objects. Distant stars seem not to move—ancient astronomers called them the fixed stars—and are thus like the bus stop; the car is like the daily rotation of the Earth, and the bicycle is like the yearly motion of the Sun.

If both these apparent motions of the Sun were regular and confined to the same plane, their angular velocities could simply be subtracted, and the length of each day would consistently stay at precisely 24 hours throughout the year. But that is not the case. The yearly wandering of the sun over the celestial sphere happens along the ecliptic, the plane of the Earth's orbit, while the daily rising and setting of all celestial objects mirrors the Earth's rotation and thus happens within the plane of the equator (Fig. 4.2). The angle between equator and ecliptic, the *obliquity of the ecliptic,* also known as the tilt of the globe (Fig. 4.3), is 23° 26' 32" for the year 2002. Through the ages, this value varies between 21° and 28°. At present, it goes down by 0.47" per year.

Thus, the Sun's apparent wandering of 360° in 365.25 days, that is, 360/365.25 = 0.985° or approximately one degree of arc per day from west to east, proceeds parallel to the equator at the solstices, but at an angle of 23.5° at the equinoxes (Fig. 4.2). Only the Sun's progress parallel to the equator counts for the length of day. Close to the equinoxes, this progress is the horizontal component of the Sun's motion, given by $0.985 \times \cos 23.5 = 0.903°$. The solar day, actually the sum of day and night, is thus longer than average at the times of the solstices, and shortest around the equinoxes.

As if that wouldn't complicate matters enough, one still has to consider the eccentricity of the Earth's orbit. According to

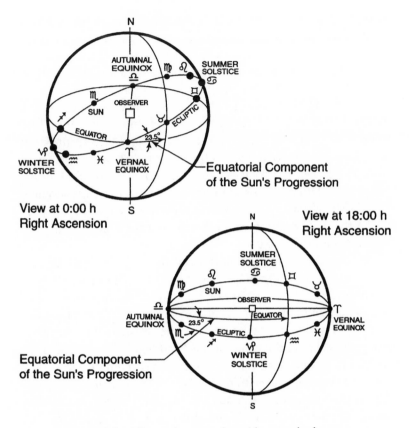

Figure 4.2 The sun's unsteady stride over the heavens

Kepler's second law, the Earth's velocity around the sun changes throughout the year, being greatest (30.3 km/sec) on January 2, when the Earth goes through the perihelion, the orbit's point closest to the sun at the distance of 147.1×10^6 km. On June 4, at the aphelion, the outermost point of its orbit (152.1×10^6 km), the Earth's orbital velocity is at its yearly low of 29.8 km/sec.

The effects are surprising: around February 12, midday comes early by 14 minutes, adding badly needed extra sunlight to short winter afternoons. Conversely, midday is late around November 3, cutting some 16 minutes off the afternoon daylight.

Figure 4.3 The tilt of the globe

The Average Solar Day

As we have seen, there are periodical changes in the solar day's length. The longest day is December 25, with 24 h 30 sec between two successive meridian transits of the Sun. Conversely, the shortest day, September 16, lasts only 23 h 59 min 39 sec. Hence, the

term "average" in the 24-hour definition of the solar day. But this is not yet the end of the story. Tidal forces, stemming from gravitational attraction of Sun and Moon, have the long-term effect of slowing down the Earth's rotation, just as tidal forces from Earth have already slowed the Moon's period of rotation to its present 27.34 days. This equals the Moon's period of revolution, which is why the Moon always shows the same face.

The Earth, eighty-one times more massive than the Moon, slows down much less. Nevertheless, 600 million years ago, when life was about to erupt on our planet, the days were $3\frac{1}{2}$ hours shorter than now. Eighty-five million years ago, as the dinosaurs waned, this difference was 20 minutes, and at present, the day's length keeps increasing at the rate of 1/600 second per century.

Contemporary Time Standards: Based on the Atom

Considering all these uncertainties, it's surprising that the definition of the second as the 86,400th part of the average solar day held until 1960, when a quantum-theoretical definition was put in place. The second became "the time of 9,192,631,770 cycles of resonance vibration of the cesium-133 atom."

The length of this time standard comes as close as it can get to the length of the traditional second, but since the absolute regularity of atomic clocks is not matched to equal accuracy by the Earth's rotation, our standard timepieces suffer occasional one-second adjustments forward or backward in order to keep them in step with day and night.

· 5 ·

Weighty Matters

Of Mass and Force

In principle, it is not a big problem if I wish to measure everything in furlongs while you want to use kilometers. Both are standard, and we simply have to work out the conversion factor, the equivalent of the exchange rate when dealing with foreign currency. In practice, however, this would lead to a flood of conversion factors from one unit to the next and thus unnecessarily complicate the mathematical description of natural phenomena.

Therefore, unit selection has been restricted to the least possible number of arbitrary standards. From these, all other units are deduced with the aim of making as many conversion factors as possible powers of ten, and of presenting the equations of physics in their simplest mathematical format.

Basic Units

In the foot, pound, second system, the set of basic units includes length (ft), mass (lb), time (sec). As we shall see, others are needed for electric current (ampere), temperature (°F), and luminous intensity (candela). From these, all other units are derived. For instance, area is measured in square foot (ft^2), volume by the cubic foot (ft^3), velocities in feet per second (ft/sec); and acceleration in feet per second squared (ft/sec^2).

The mass of the pound (lb) is a basic unit in the United States, and the pound as unit of force (lbf) is a derived unit, defined as

weight. More precisely, it is the gravitational force between a one-pound mass and the mass of the Earth at a place at which g = 32.174 ft/sec^2 is the acceleration due to gravity. At such a spot, a weight lifter hoisting a barbell of 200 lb of mass exerts the force of exactly 200 lbf.

Mass and Force

Although the pound of mass and the pound of force coincide numerically in our daily environment, this would not be the case if the weight lifter took his barbells to the Moon, where gravity is only one-sixth that on Earth. On the Moon, the mass of the barbells is still 200 lb, but the force needed to lift them becomes a mere 200/6 = 33.3 lbf. The same effort needed to lift 200 lb on Earth would allow the athlete to handle a whopping 1200 lb on the Moon.

Mass and weight must therefore be expressed in different units, although gravity on Earth is uniform enough to make us forget such details in the course of daily life. These units are the pound of mass (lb) and the pound of force (lbf); the kilogram of mass (kg) and the kilogram of force (kgf); and the newton (N), a unit of force named after Sir Isaac Newton of falling-apple fame. Derived units make the distinction important. These include lbf·ft for torque, ft·lb or N·m (joule) for work, but lb/ft^3 or kg/m^3 for density, among others.

How to Derive a Unit

If a unit's derivation is not obvious, it usually comes from the laws of physics. Take acceleration, for example. For a bar pendulum, the oscillation period is $T = 2\pi \times \sqrt{L/3g}$. Reshuffled, this gives for gravity acceleration:

$$g = \frac{4\pi^2}{3} \times \frac{L}{T^2}.$$

Since the dimensionless factor $4\pi^2/3$ can be neglected in unit-related calculations, the unit for acceleration follows from L/T^2 = m/sec^2 as meter per second squared, or m · sec^{-2}.

Therefore, acceleration per hour is not 3600 times the acceleration per second, but it is $3600^2 = 12.96 \times 10^6$ times greater. Note that the sexagesimal subdivision of the hour is the reason why, even in metric units, this conversion factor is not a round figure. Had the French metric clock (Fig. 4.1, Chapter 4) of two centuries ago been accepted, this factor would be a straight $(100 \times 100)^2 = 10^8$.

As another example of deriving a unit, take a wooden beam of width b and height h, typified by the popular "two by four." Its load-carrying capacity is given by its section modulus $bh^2/6$. From this, the dimensions of the section modulus are evaluated as $m \times m^2 = m^3$, or, in FPS units, $ft \times ft^2 = ft^3$. Although these units are identical to those for volume, section modulus and volume are otherwise unrelated.

In metric units, section modulus is most conveniently expressed in cm^3 (cubic centimeters) for most practical applications. Since one meter equals 100 centimeters, the conversion factor becomes $100^3 = 10^6$, or one million. For comparison, converting a section modulus from cubic foot to cubic inches implies multiplication with $12^3 = 1728$, which shows again the advantage of 10-based units throughout, offered by the metric system and the International System of Units (SI).

While length, mass, and time are the natural choices for basic units, here again there is no objective rule as to which units are basic and which are derived. For instance, could we make a unit we instinctively consider as derived, such as the density of water, the basic unit in a yet to be defined system of measures and deduce the unit of length? Calling the new basic unit of density "Den" and the new length unit derived from the Den the "Ankle," abbreviated Ak, the definition of density D as the fraction mass/volume, $D = m/V$, rearranged into $V = m/D$ gives dimensionally

$$Ak^3 = \frac{lb}{Den} \text{ and } Ak = \sqrt[3]{\frac{lb}{Den}}.$$

From the weight of one cubic foot of water, $62.428\ lb/ft^3$, this formula yields for the length of the Ankle

$$1 \text{ Ak} = \sqrt[3]{\frac{1}{62.428}} = 0.2521 \text{ ft.}$$

The density-derived unit, the Ankle, would thus measure about one quarter of our present foot, or about 3 inches; its dimension, however would be an awkward $(\text{lb} \times \text{Den}^{-1})^{\frac{1}{3}}$.

Skeptics can check this result by figuring the volume of a cube of one-Ankle side, and multiply the result with the density of water. With 0.2521 conventional feet to the new foot, the cube's volume becomes $0.2521^3 = 0.01602 \text{ ft}^3$. Multiplied with the water's specific gravity of 62.428 lb/ft^3, this yields the cube's mass as $0.01602 \times 62.428 = 1.000$ lb, in perfect concordance with the definition of the hypothetical basic unit of the Den (Fig. 5.1).

While density as a basic unit would be everything but desirable, it would function correctly with all the laws of physics, although dimensions in that kind of system would look somewhat strange, to say the least. Velocity would be measured in $\text{lb}^{\frac{1}{3}} \times \text{Den}^{-\frac{1}{3}} \times \text{sec}^{-1}$, and acceleration in $\text{lb}^{\frac{1}{3}} \times \text{Den}^{-\frac{1}{3}} \times \text{sec}^{-2}$. But conversely, volumes and section moduli would get the simple unit lb/Den.

Farfetched as all that might seem, it is not without real life parallels. The density of water in both the metric and the SI system is 1 kg/dm^3 (Fig. 5.2), but the originators of the metric system

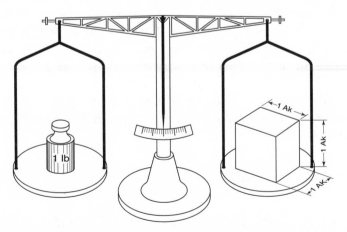

Figure 5.1 The "ankle" (Ak), derived from the density of water. Example of an inverse approach.

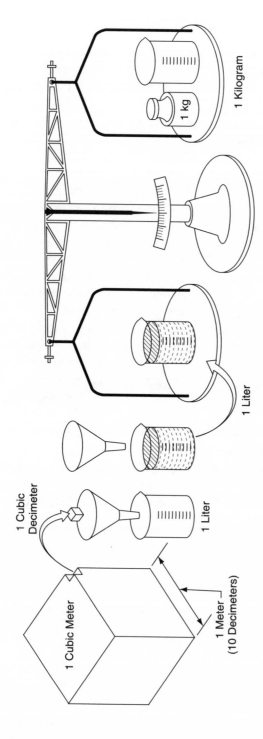

Figure 5.2 From cubic decimeter to liter, and from there to kilogram

achieved this handy condition through a more direct approach: they started with the length of the meter and subsequently standardized the weight of the kilogram to be equal to the weight of one cubic decimeter of water.

However, the International System's standard unit is based on the meter. From the relation 1 m = 10 dm, we get 10^3 dm^3 = 1000 dm^3 = 1 m^3, so that for all practical purposes, the density of water is 1000 kg/m^3, or one metric ton per cubic meter.

The Absolute System of Weights and Measures, forerunner of the metric system, is based on the units centimeter/gram/second, rather than meter/kilogram/second, and thus uses 1 g/cm^3 for the density of water. Since one decimeter divides into ten centimeters, and therefore 1 dm^3 = 10^3 = 1000 cm^3, while the kilogram equals 1000 gram, 1 g/cm^3 is the equivalent of 1 kg/dm^3.

Here again, the search for the most handy sets of units rather than some law of physics leads to the selection of length, mass, and time as the basic units of our Customary System, the Absolute System, and the International System of Units. By contrast, the Technical System, forerunner of the SI, has length, force, and time, as basic units. To sum up and putting the ampere, kelvin, candela, and mole aside, the *basic units* are as follows:

In the *International System:* meter (m), kilogram of mass (kg), and the second (s).

In the *Technical System:* meter, kilogram of force (kgf), and the second (s).

In the *FPS System:* foot (ft), pound of mass (lb), and second (s).

· 6 ·

Gravimetric Standards

Because we instinctively equate the weight of an object with the force it takes to lift it, weight standards, or weight-based force standards, have been part of systems of measure throughout history. Only the development of precision instrumentation for gravity measurements made people aware that the weight of a given mass *did* vary with its location on Earth. In Madrid, Spain, gravitational acceleration g is 9.79981 m/sec^2, while in Greenwich, England, it is 9.81188 m/sec^2, a difference of 0.123%. Some people think that Bob Beaman's record-breaking long jump at the Mexico Olympics—a record that would stand for over 25 years— was possible because Mexico City has a relatively low value of g.

Such inconsistencies of gravity around the globe could have been known much earlier if laboratory-grade analytical scales with typically 0.001 gram precision could show them, but they couldn't. As the object in one of the scale's trays gets heavier or lighter because of gravity, so do the weights placed on the other tray (Fig. 6.1). This means the traditional beam scale (including the masterpiece of metrology in Figure 6.2) does *not* detect variations of gravity at all. By contrast, spring scales, including industrial load cells or their popular kitchen variety, fully react to changes in gravity. On the planet Mars or on the Moon (Fig. 6.1), their readouts indicate the size of the gravitational force on these celestial bodies relative to gravity on Earth.

Placed at rest at the altitude where the space shuttle orbits, some 122 nautical miles (226 km) above the Earth, a spring scale would show the weight of one pound of mass as approximately 0.93 pounds. Thus, the sensation of weightlessness that Shuttle

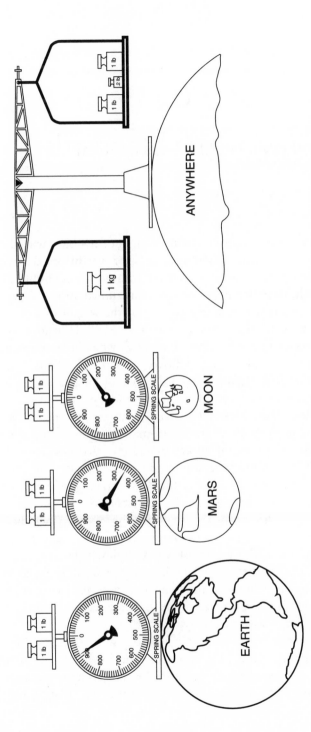

Figure 6.1 Spring scales show gravity variations; beam scales don't. A spring scale measures gravitational force and, thus, gives different readings on locations with disparate gravities. A beam scale compares the masses resting on its two trays and, thus, operates independently of prevailing gravitational forces.

astronauts experience is not the microgravity of interplanetary space, but the result of centrifugal forces generated by the Shuttle's motion around the Earth, which opposes and seemingly annuls terrestrial gravity.

Gravimetry

If spring scales had the necessary sensitivity, they could show gravity variations from place to place. But they do not. Instead, precision instruments, called gravimeters, must be used. They measure gravitational force indirectly, either by monitoring the free fall of calibrated rods in a vacuum or from the period of a bar pendulum, which, according to the formula $T = 2\pi \sqrt{L/3g}$ (Eq. 2.1 in Chapter 2), changes inversely with the square root of gravitational acceleration g. Modern gravimeters are sensitive enough to detect gravity variations of one hundred millionth of g.

The International System does not recognize any separate unit for gravitational force other than the newton and its multiples. However, the Gal, named after Galileo Galilei, of 1 cm/sec^2, is frequently used. Standard gravitational acceleration might thus be expressed as 980.665 Gal.

Because of the oblate shape of the globe, the Earth's radius at the poles is shorter by 21 km (13.3 miles) than at the equator. This alone makes for a difference of 1.8 Gal in gravitational force, while centrifugal forces, stemming from the Earth's rotation, add another 3.4 gal. Altogether, objects are thus 5.2 Gal, or 5.2 × 100/980.665 = 0.53% heavier at the poles than around the equator. A 190-pound person would lose an entire pound while traveling from the Arctic to the tropics, something dedicated weight watchers should remember in their vacation plans.

Lunar Olympics

Space travel has provided an opportunity to directly experience phenomena that used to be the domain of theoreticians alone, such as the size of gravity on neighboring celestial bodies, or the absence of gravitational forces altogether. With the Moon's mass 1/81 that of Earth, gravity there would be 1/81 g if Moon and

Earth were of the same size. But the Moon's radius is only 1740 km, compared to the Earth's 6375 km, which makes astronauts on the Moon stand $6375/1740 = 3.66$ times closer to its center than their peers back at NASA stand to the center of Earth. This, according to Newton's "inverse square law," boosts gravitational force on the Moon's surface by a factor of 3.66^2, which makes objects on the Moon weigh $3.66^2/81 = 1/6$ of their weight at home.

Likewise, with the mass of planet Mars $1/9.35$ that of Earth and its radius 0.532 earth radii, a one-pound object on the surface of Mars weighs $1/(0.532^2 \times 9.35) = 0.38$ lb.

The true mark of lifting the human body is in the category "jump above own head," where Franklin Jacobs in 1978 set the record at the modest height of less than 2 feet. Thus, even if astronauts didn't have to wear their heavy spacesuits on the moon, they couldn't jump much beyond $6 \times 2 = 12$ ft. Evidently, jumping is not the humans' strongest suit. The common flea has the capacity to jump 130 times its own height; a lunar flea could leap from the Moon's surface to a height equivalent to that of the observation deck of the Empire State Building.

In other sports, however, extraterrestrial locations would allow for impressive achievements. A tennis player on the Moon would have to play on a court the size of a soccer field, otherwise all shots would be "out." On the Moon, as on Earth, the ball follows a straight line in the direction you hit it while simultaneously dropping vertically downward, under the action of gravity. Thus, the familiar curved flight path results—a parabola. As gravity is less, the ball drops downward more slowly and so Pete Sampras's serve could never bounce within the lines.

Baseball might be less interesting on the Moon. Sinkers, sliders, and knuckle balls wouldn't work, so every pitch would be a fastball. With no air resistance, the pitcher would easily clock at 100 miles per hour.

The golf ball that astronaut Alan Shepard hit in 1971 during the Apollo 14 mission on the Moon, must have followed an almost perfectly parabolic flight path. After a tentative shot at a small crater 100 feet away, Shepard hit pure while shouting "here it goes!" into his helmet's microphone. His bulky spacesuit kept him from using both hands on his collapsible utility pole that had

a six-iron attached, so Shepard could not hit his ball as well as an Earth-based player. Even if Shepard could only hit a ball 150 yards on Earth, a similar shot on the Moon would go some $6 \times 150 =$ 900 yd, or about 0.51 mile, beating all earthly records by nearly 500 yards. If that falls short of the "miles and miles" we remember from speeches in the late Shepard's honor, it still sounds like many a golfer's dream.

A Gravitational Yardstick

While the acceleration of a speeding car is easily checked by noting how much the needle of its speedometer climbs per second, gravitational acceleration g is historically measured from free fall. The equations are

$$s = \frac{g}{2} \times t^2 \tag{6.1}$$

or

$$g = \frac{2s}{t^2}, \tag{6.2}$$

where s is the distance covered by the falling object in the time t. If Galileo had dropped a cannonball simultaneously with a wooden sphere from the 179-foot-high leaning tower of Pisa, he would have found that both hit the ground simultaneously 3.34 seconds later. This would give gravitational acceleration at the city of Pisa as

$$g = \frac{2s}{t^2} = \frac{2 \times 179}{3.34^2} = 32.1 \text{ ft/sec}^2.$$

With blatant disregard for the Aristotelian philosophy of his times, Galileo interpreted such experiments as proof that the acceleration (velocity gain per second) of a free-falling object was independent of its mass. This gave birth to one of the basic laws of kinematics.

That said, Galileo must have been aware that a cotton ball or one made of crumpled paper actually takes longer to descend than a heavier object. As we now know, even a skydiver hardly accelerates beyond 200 kilometers per hour. This is due to the drag force of the atmosphere. Still, Galileo's great power of abstraction led him to strip this phenomenon of all secondary influences and get down to the governing principles behind it.

Four centuries passed before a spectacular experimental proof of Galileo's contentions could be staged. Astronaut David Scott of the Apollo 15 mission dropped a feather along with a hammer on the lunar surface. As the Moon has no atmosphere, both objects, true to Galilean principles, reached the ground at the same instant.

Impressive as such demonstrations might be, they should not surprise anyone these days. The laws of gravitation and inertia allow us to predict the following results.

We set m_1 and m_2 as the masses of Earth and of the falling body, respectively. R is the earth-radius, g is the gravitational acceleration, and G is the gravitational constant. The universal law of gravitation says that the force F on the falling object is given by $F = G \times m_1 m_2 / R^2$. This, however, must equal the weight, W, of the object, which is $W = m_2 g$. Thus, we get $G \times m_1 m_2 / R^2 = m_2 g$. Canceling m_2, this becomes $g = G \times m_1 / R^2$, a term that no longer contains m_2, the mass of the falling body. Thus, we get a mathematical confirmation of what Galileo Galilei postulated long ago.

Standard Gravitational Acceleration

In talking about the force of gravity on other planets, or the acceleration on a roller coaster, g is the obvious unit. The *average terrestrial gravitational acceleration* has been standardized at $g = 9.80665$ m/sec^2, that is, $g = 32.17405$ ft/sec^2. Gravity, in general, is expressed as a fraction or multiple of that standard. On the Moon, it is $g/6$, on Mars $0.38\,g$, and on Jupiter $2.6\,g$. Back home, Jupiter's gravitational powers have been beaten by Ron Toomer's famous roller coaster designs such as the Scream Machine and the Magnum XL-200, which, by way of inertial forces, occasionally submit their riders to $3.5\,g$. Astronauts fly aboard the "Vomit

Comet," which subjects them to similar accelerations. At 4 g or more, there's a good chance of passing out.

Standard gravity exists at any place at sea level and at 45° of latitude, save for minor variations induced by the presence of mountain ranges, mineral deposits, and other irregularities in the inner and outer structures of the globe. With the value for g defined, weight is understood as the gravitational attraction at a place with standard gravity acceleration of g = 9.80665 m/sec^2.

This exact value makes a weight standard feasible, but its definition would call for not one, but two characteristic figures: standard mass and standard gravity acceleration. If, for instance, our imperial standard pound would be made the standard of weight, the condition "at a place with standard gravity acceleration of 9.80665 m/sec^2" would have to be appended. This need for dual definition caused weight standards—although historically they were a natural—to be excluded from all of the present systems of weights and measures, while mass standards prevailed.

Figure 6.2 Precision balance (2500 pounds) with mass standards, given to the state of Maryland in 1970 with a complete set of metric and customary standards for mass, length, and volume

· 7 ·

The Matter with Mass

A body's mass can be perceived either through its *weight,* or through its *inertia,* that is, a body's resistance to a change of its state of motion. In the first case, we are talking about gravitational mass, in the second about inertial mass.

An aircraft uses part of its engines' power to keep its own gravitational mass from falling back to the ground (Fig. 7.1), but a dragster uses its engine to overcome the reluctance to a change of motion of its inertial mass (Fig. 7.2).

Newton's Estimates

Isaac Newton, eternalized as the discoverer of gravity, was privately known for his intensity of involvement in the problems of physics, which often made him miss sleep—and even his meals. At one such occasion, the story goes, a worried housekeeper entered his workplace carrying an egg, a pot of water, and a pocket watch and instructed him to heat the water on his stove and subsequently boil the egg for exactly 3 minutes. Returning some time later, she found her master posted at the stove, staring fixedly at the egg he held in his hand, while the watch was boiling wildly in the pot.

Such obsession with his work must have bestowed Newton with a kind of sixth sense for the laws of nature and their interpretation. For practical application of his "inverse square law" $F = G \times m_1 m_2 / R^2$, Newton needed the value of the gravitational constant, G. Though it can be found by applying Newton's law to a free-falling object, and get $G = gR^2/m_1$, Newton had no data

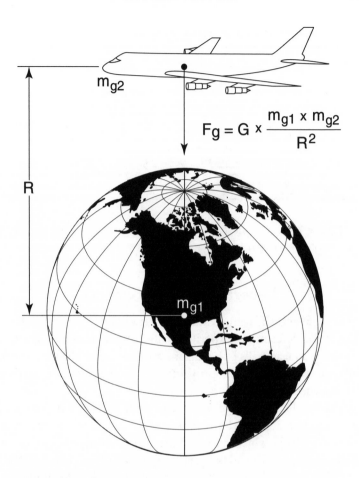

$$F_g = G \times \frac{m_{g1} \times m_{g2}}{R^2}$$

Figure 7.1 Gravitational attraction between the mass of Earth and an aircraft

on m_1, the mass of the Earth. Since the size of our planet was pretty well known at those times, Newton could estimate the Earth's density in order to figure its mass from its size. His guess must have been fairly accurate, since the value he came up with for the gravitational constant, $6 \times 10^{-11} \text{N·m}^2/\text{kg}^2$, lies within 10 percent of the correct figure. However, he never knew of his good luck, since it was not until 1798 that a direct measurement of gravitational attraction was successful.

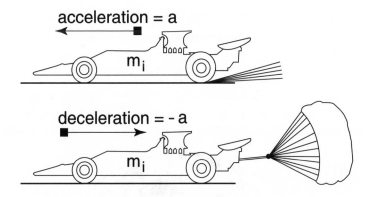

Figure 7.2 The inertial mass of the dragster opposes its acceleration (speedup), as well as its deceleration (slowdown).

Laboratory Measurements

Henry Cavendish, using Coulomb's traditional torsion balance (Fig. 7.3), determined experimentally that the mutual attraction of two small lead balls by a pair of stationary, much heavier spherical chunks of lead was proportional to the product of their masses, according to $F \propto m_1 \times m_2$.

The smaller balls were mounted on the ends of a horizontally swiveling beam, leveled at the height of the center of gravity of the heavier balls, which rested on the floor. Inverting the positions of the small and big lead balls relative to each other (Fig. 7.3, steps 1 and 2) makes the torsion balance turn by twice the angle corresponding to their mutual gravitational attraction.

This procedure not only doubles the angle of the beam's deviation, but most importantly it eliminates the uncertainty of the zero position, a feature inherent to torsional counterforce instrumentation. Industrial instruments with spring return of the needle, like voltmeters and ammeters, solve this problem by using a pair of coiled leaf springs, wound clockwise and counterclockwise, respectively, instead of a single one.

The Hungarian physicist Roland von Eötvös (1848–1919) improved the accuracy of the torsion balance to the point that even the dependence of gravitational force from the distance of

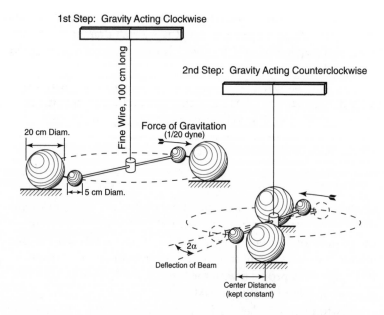

Figure 7.3 Torsion balance for measurements of gravitational force (data from Cavendish's setup in 1798)

the mutually attracting objects (the inverse square law) could be directly checked and proven. Further perfection was reached by Princeton physicist Robert H. Dicke, who demonstrated the equivalence of gravitational and inertial forces to an accuracy of one part in 10^8. Dicke, widely known for his role in the development of microwave radar, became involved in a quest to test Einstein's general theory of relativity. This led to his high-precision measurements on gravitational and inertial mass equivalence.

Inertial Force

If nature's agents had seats in Congress, inertia would be the most conservative among conservatives. Inertial forces never drive, they only oppose change: *change* from the state of rest to the state of motion, such as a car speeding away from a parking lot; *change* of motion, such as when we accelerate or brake; and *change* in the

direction of motion while speeding through a narrow bend. As the devilish Mephisto says in Goethe's *Faust* drama: "I am the spirit that negates all and forever." If inertia could talk, it could have made such a statement.

Not surprisingly, inertial forces grow in proportion with mass, *m*, and, likewise, with the rate of change of velocity—that is, acceleration or deceleration, respectively. The inertial force *F* in Newton's second law of motion is expressed as

$$F = m \times a. \tag{7.1}$$

(Pedants will know that this equation holds good only for systems of constant mass.) For instance, a 2000 kg dragster, built to reach the velocity of 144 km/h or $144 \times 1000/(60 \times 60) = 40$ m/sec in a mere 4 seconds, will accelerate at the rate of $40/4 = 10$ m/sec. Its engine must propel it with the force of $F = 2000 \times 10 = 20,000$ newton or $20,000/9.807 = 2040$ kgf.

Newton's second law leads to an absolute unit force, one that accelerates a mass of 1 kg by 1 m/sec². If that sounds abstract, think of a free-falling object, driven by the force of its own weight. It accelerates by $g = 9.80665$ m/sec². To make it accelerate at a mere 1 m/sec², we'll need only 1/9.80665 of that force. That fraction became the newton (N). As with the kilogram of mass, the newton of force is an absolute unit. Unlike the older weight-based units, it is easily reproducible in any part of the cosmos under any gravitational conditions. But since we Earthlings use scales rather than accelerometers to compare masses with one another, we may picture the magnitude of the newton as comparable to the weight of $1/9.80665 = 0.10197$ kg.

With $a = \Delta v/\Delta t$, Newton's second law reads $F = m \times \Delta v/\Delta t$, and in this version, it helps us understand many aspects of daily life. The impact of a car on a solid wall reduces the vehicle's speed from *v* to zero. That makes $\Delta v = v - 0 = v$. It does this in no time at all ($\Delta t = 0$) so that Newton's equation results in $F = mv/0 = \infty$. That's why you should wear a seat belt and have a car with air bags and crumple zones. That way, if you hit a wall, at least the passengers will come to a gradual stop.

Most important, Newton's second law, reshuffled into $F \times \Delta t = m \times \Delta v$, provides two new concepts in physics, *impulse and*

momentum, showing up on the left and right sides of the equation, respectively. In the absence of external forces, $\Delta(mv) = 0$, makes the momentum a constant. Suppose a Doomsday asteroid of mass $1/100$ that of the Earth is on course to hit us. Before the collision, it has speed v_{ast}, but the Earth is at rest (relatively speaking). Afterwards, the Earth and asteroid are stuck together and move at speed v. Then $m_{ast}v_{ast} + m_{earth}v_{earth} = (m_{ast} + m_{earth})v$, and with $v_{earth} = 0$:

$$v = \left(\frac{m_{ast}}{m_{ast}m_{earth}} \right) \times v_{ast}.$$

If a body the size of the Moon ($1/81$ the mass of Earth) would fall from a great distance straight into the Earth, its impact velocity would equal escape velocity of 11 km/sec. For that case, the above equation (somewhat reshuffled) yields

$$\frac{v}{v_{ast}} = \left(\frac{m_{ast}}{m_{ast}m_{earth}} \right) = \frac{\dfrac{1}{81}}{1 + \dfrac{1}{81}} = 0.012, \text{ slightly above 1\%!}$$

Thus, the effect on the Earth's motion would be minimal, but according to $E = mv^2/2$ for kinetic energy, the devastation proportional to the square of the impact velocity of typically 11000 m/sec, would be immeasurable.

Another way of demonstrating impulse and momentum is shown in Fig. 7.4A–D by the puzzling behavior of ideally elastic (make that "almost") bodies when they impact each other. In Figure 7.4A, a pair of hardened steel balls, such as ball bearings, are mounted on strings like pendulums. As the ball on the right reaches velocity v at its lowest point, its momentum becomes $m \times v$. On impact, both balls, the one at rest and the one that swings, deform elastically and then, in the process of retaking their former shape, recoil from each other. The ball on the right obtains the momentum $- m \times v$, which brings its total momentum to $m \times v - m \times v = 0$. That is, it stops dead. By contrast, the ball on the left gains the momentum $m \times v$, which makes it swing to the left. Thus, the balls exchange their motion with every swing, but the sum of their momenta remains constant all the time.

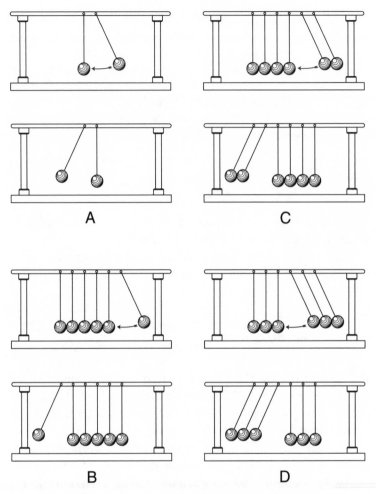

Figure 7.4 Momentum and impulse demonstrations

That works just as well with several balls in between, as in Figure 7.4B. More amazingly still, if we play the game lifting two or three balls simultaneously, two or three balls will respectively swing upwards on the other side (Fig. 7.4C and D).

Since inertial and gravitational forces make part of our daily life experience, we tend to take them for granted. Originally, gravitation was thought of as comparable to electrostatic or magnetic

attraction, while inertia has been seen as a property inherent to matter itself. Einstein's general theory of relativity brings both on a common denominator—so to speak—by use of a four-dimensional nonlinear system of coordinates. The illusive "grand unified theory" would extend that system to include electrostatic and magnetic forces, but that would still leave us with explaining the strong and the weak nuclear interactions. The question here seems to be how far back we can follow the law of action and reaction. In mathematics, the necessity of a set of unproven axioms has been accepted. Whether or not such a set should exist in physics has remained subject to discussion.

· 8 ·

Empire of Light

Luminosity and Intensity

Ancient Egyptians worshiped in the Sun the figure of their god Ra, the power that maintained all life on Earth. Without sunlight, how long could humans endure? And, given the choice between a world of riches and a ray of sunlight, Diogenes, the Greek philosopher of the fourth century B.C., chose the latter. Convinced that material possessions were unlikely to bring happiness, Diogenes lived the life of a hermit on the outskirts of Athens, using nothing better than an old barrel to protect himself from the elements. He was a familiar figure to Athenians from his occasional trips to the city's crowded marketplace, where he appeared stark naked, waving a lit lantern in bright daylight. When asked what in the world he was trying to find with his lantern, he lacomically replied, "Humans . . . humans."

Such tales caught the attention of Alexander the Great, then the ruler of the ancient world and beyond. Trying to make the philosopher contradict his own teachings, Alexander appeared one morning in front of Diogenes's barrel house to announce that "anything in the world Diogenes could wish for" would be his. Diogenes, probably warming up in the rays of the morning sun after a night without blankets, mumbled in response, "I wish you wouldn't block the sun."

This story may seem silly by the socioeconomic standards of our time, but it is right on the money in an absolute system of values. Light rays have taught us the size and the age of the world,

and most important, we came to know the velocity of the propagation of light in vacuum as the one and only invariant magnitude in the cosmos. This knowledge allowed Einstein to reformulate the laws of nature with his special theory of relativity. Light, once considered as one of many natural phenomena, then became a pillar of physics.

The Puzzling Nature of Light

Light is a form of electromagnetic radiation. So too are radio, television, and radar signals, but they have lower frequencies and thus longer wavelengths. Microwaves, which fill the universe and are left over from the Big Bang, are also a form of radiation. X-rays, γ-rays, and ultraviolet light (which can be seen by cats and which caused white shirts to glow in 1970s discos) have shorter wavelengths.

Frequencies are traditionally measured in hertz, a unit that counts the number of oscillations per second and was named after the German physicist Heinrich Hertz. Hertz proved that electricity can be transmitted in electromagnetic waves, then fittingly called "Hertzian waves," which travel at the speed of light and share most of the other properties of light. His experiments opened the doors for the development of the wireless telegraph and radio.

The broad spectrum of radio waves spans from kilometers of wavelength down to centimeter and millimeter waves. Collector's items among table radios may still show, along with the regular scales for broadcast and short waves, a third scale, marked "long waves," covering lower frequencies. In the very early days of radio, frequencies above the 1660 kHz upper limit of the broadcast band were considered unfit for professional communication and got freely distributed among amateurs, later tagged as HAM operators. What a bargain in the light of the 100 billion German marks (about 40 billion dollars) European industries spent in the year 2000 alone on a handful of far shorter wavelengths for hand-held telephones.

In the more recent past, kilometer waves have been found useful for underwater detection of hostile submarines, to the dismay

of environmentalists, who fear possible bad health effects on sea creatures.

Light waves are much shorter than even the shortest radio waves, and are traditionally measured in angstrom units (named after the Swedish scientist Anders Jonas Ångström). Ten million of his units fit into one millimeter. However, the International System of Weights and Measures (SI) has made the ten times greater unit of nanometer (1 nm = 10^{-9} meter) official. Light waves from the Sun measure 555 nm or 5550 angstroms at the wavelength of highest intensity.

For some purposes, light can be reflected, refracted, and diffracted just like waves. But here, too, light has some surprises in store for us. The greatest astronomical telescopes of the world operate on the principle that a parabolic mirror reflects all incident light (parallel to its axis) into one single focal point. Legend has it that in ancient Greece Archimedes fought off an invading force from Syracuse by using concave bronze mirrors to focus the rays of the sun on the vessels' wooden hulls and setting them ablaze. "Se non é vero, é bene trovato" (if not true, it's nicely thought out), as the Italians used to say. In your days of scouting you probably learned to use a magnifying glass to start a campfire, and sometimes even to simulate the sting of a bee on the skin of an unsuspecting sunbather. But the image of the Sun, projected at, say, a 100-meter (328 feet) distant target, would measure about 3 feet in diameter. Even the light-gathering power of a mirror the size of the Hubble Space Telescope would hardly suffice to heat an area of that size to the point of ignition of damp timber.

While reflection works along the lines of rubber balls ricocheting from a wall, refraction is linked to the velocity of light within the refracting material. In crossing a piece of windowpane with a refractive index of, say, $n = 1.53$, light travels 1.53 times slower than in vacuum, which is apparent when you hold a piece of glass between the objective lens of a microscope and the probe the instrument is focused on. Inserting the glass moves the image out of focus. For refocusing, you have to crank the eyepiece somewhat down until you reach the point where the time the light takes to travel from the probe to the objective lens of the microscope is what it was without the glass in between. The distance between

probe and objective lens has changed, but the time light traveled that distance has remained constant.

Light picks the fastest rather than the shortest way from point A to point B, just like a person walking over uneven terrain would seek out the fastest track. For instance, imagine, yourself walking daily from your tent to the beach and swimming out to your sailboat (Fig. 8.1). Since you walk faster (75 yards per minute) than you swim (50 yards per minute), you may try out a number of alternatives to get to your boat, such as A and B in the illustration. But the shortest path in terms of time is in C. For proof, take the space you walk as $s_1^2 = x^2 + y_1^2$, and the space you swim as $s_2^2 = (d - x)^2 + y_2^2$. The time you are under way, $T = s_1/v_1 + s_2/v_2$, is shortest with dT/dx equal to zero, which leads to the equation: $v_1^2 \times [1 + y_1^2/x^2] = v_2^2 \times [1 + y_2^2/(d - x)^2]$. Substitute y_1/x with tan α and $y_2/(d - x)$ with tan β, and use the identity $1 + 1/\tan^2 \alpha = 1/\sin^2 \alpha$, and you get sin $\alpha/\sin \beta = v_1/v_2 = n$.

"So what?" you might be tempted to ask after all that math. Well, here we have an amazing parallel of daily life experience and the complexities of wave physics; your tracking and swimming got us the *Law of Refraction*, pure and simple. The enticing photographs of our children and grandchildren turn out because light from these "objects" is in such a great hurry to get through the camera's lenses to the plane of the film. And the image we see through binoculars is there because light tries to get through the instrument in the shortest possible time. The word "lens" comes from the Latin word for lentil, the vegetable with its characteristically curved shape, akin of a convex lens. The laws of physics are simple to understand unless you make them complicated. Just like the boy in a hospital, who asks another boy if he belongs to internal medicine or surgery and, getting a blank stare for an answer, rephrases his question into: "Were you ill when you came in, or did they make you ill afterward?"

In many cases—such as in the photoelectric effect, for which Einstein was awarded the Nobel Prize—light behaves just like particles, in other cases like waves. This so-called duality in the behavior of light caused great confusion among scientists in the early 1900s who preferred to explain all natural phenomena by analogies with mechanical models. That the ringing of an electric bell,

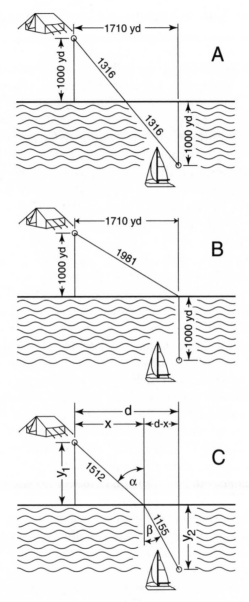

Figure 8.1 Index of refraction and the velocity of light

encased in a vacuum chamber, waned when the air was pumped out, was taken as positive proof that waves needed a carrier for their propagation: the atmosphere for soundwaves on Earth, and the "world ether"—an invisible, ideally thin and penetrable sort of matter—in space. This concept survived until 1881, when the famous Michelson-Morley experiment found no differences in the speed of light when measured in or against the direction of the Earth's orbital motion through the hypothetical ether, or at right angles to it. The experiment disproved the ether's very existence. Most significant however, the experiment heralded a new era in physics, hinting that the speed of light is constant and invariable throughout the cosmos.

How Much Light from Yonder Window Breaks?

Standard light sources were around long before the age of electric illumination: the Hefnerkerze (Hefner candle) was used in continental Europe, and "candle power" was used in Great Britain and the United States. The Hefnerkerze was essentially an oil lamp of closely specified dimensions, fueled by amylacetate, an oil, obtained from pears. The United States' "candle power of luminous intensity" was defined as the light of a spermaceti wax candle (made from sperm whales) consuming itself at the rate of 120 grains per hour. That nonmetric unit, the pound, equals 7000 grains.

A Law of Universal Significance

If a 1000-watt lamp properly illuminated the paintings on the walls of an art museum, a room twice as long, deep, and high would need $2^2 \times 1000 = 4000$ watt for the same degree of illumination. Likewise, a hall three times as big would need $3^2 = 9$ lamps, because the power needed for illumination increases with the total surface area. Inversely, sticking with the original single lamp would reduce the paintings' illumination to one-ninth.

This demonstrates the *inverse square law of optics*, which states that the intensity of illumination from a light source decreases

proportionally to the square of the distance from that source. This is much like the force of gravitation, which decreases with the square of the distance between mutually attracting masses.

Photometry: The Science of Measuring Light

The German scientist Robert Wilhelm Bunsen (1811–1899) is probably best remembered for his Bunsen burner. But he also invented the "grease spot photometer," which is remarkable for its ingenious simplicity. It enabled scientists to use the inverse square law of luminosity to examine the power of a light source long before electronic photometers came into existence. In Figure 8.2, a standard candle and the light source being tested (the test-source, represented by a group of three candles), are placed before and behind a screen of white paper with an oil stain in its center. The test source is moved back and forth until the blot can't be made out on the screen. Too close, and the stain appears

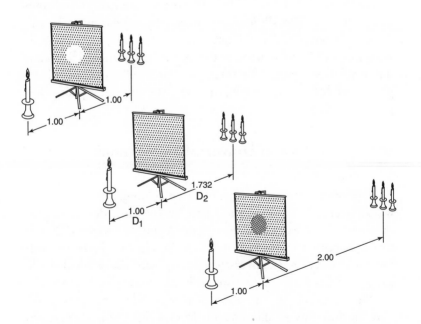

Figure 8.2 Photometry and the inverse square law of luminance

bright; too far, and the stain appears dark. When the stain vanishes, the illumination from the test source and from the standard candle has to be the same. All you need to do then is measure the distance of the two light sources from the screen.

With L_1 for the luminosity of the standard candle, L_2 for the test source, and D_1 and D_2 for their respective distances from the screen, the illumination of the screen from the front is proportional to L_1/D_1^2, and from the rear it is proportional to L_2/D_2^2. As these must be equal (because the stain has disappeared),

$$\frac{L_1}{D_1^2} = \frac{L_2}{D_2^2}.$$

Rearranged, this gives the formulas

$$\frac{L_2}{L_1} = \left(\frac{D_2}{D_1}\right)^2 \text{ or } \frac{D_2}{D_1} = \sqrt{\frac{L_2}{L_1}}.$$

In our case, $D_1 = 1$ and $D_2 = \sqrt{3} = 1.732$. So as $L_2 = (D_2/D_1)^2 L_1$, we know that $L_2 = 1.732^2\, L_1$, so that the test light source has a luminosity of three standard candles.

Candela: The New Candle Power

Spermaceti wax candles are fairly simple. Electric lightbulbs are not, for there is no simple relation between the luminous intensity of a lightbulb and its wattage. The hotter the filament, the more light one gets for the watt, so much so that in the early days of electrification, lightbulbs were sold by candle power (i.e., intensity) rather than by wattage.

Although lightbulbs with precision filaments and closely defined operating voltage have become secondary standards after all, an updated unit of luminous intensity, the *candela,* was defined in 1940 as a $1/60$ cm^2 area of black-body radiation at the solidification temperature of liquid platinum (1769 °C). The odd figure of $1/60$ cm^2 ($1/387$ in^2) was selected because it brought the new unit close enough to the classical candle power as to keep the traditional tables of photometric quantities unchanged.

The term *black-body radiation* for the light from a batch of molten platinum suggests a setup where radiation occurs without loss of energy. Ideally, a black body absorbs all incident radiant energy but also emits as radiation all the energy commensurate with its temperature. In a real-life scenario, a black car body absorbs only about 95 percent of the sunlight it gets, and heats up until it radiates as much energy into the environment as it absorbs. At the other end of the scale, clean metal surfaces absorb only 4 to 6 percent of the incident radiation.

The tips of the electrodes of carbon-arc floodlights need to heat up to about 4000 °C in order to generate as much radiation as an ideal black body would produce at 3500 °C, which brings us to the concept of "black-body temperature" as an alternative measure of a material's emissivity. A star, for example, can have its luminosity measured with an instrument called a *bolometer*. Assuming the star behaves like a black-body radiator, one can assign it a "black-body temperature" and get a rough idea of how hot it really is.

Although a black body as such has no counterpart in nature, a closed box with a small opening centered on one of its walls can simulate it. The hole always appears pitch dark, even if the inner walls of the box were (tentatively) painted white. This because multiple reflections inside cause the almost complete absorption of what little radiation enters through the hole, and the incoming rays have virtually no way of leaving. The *candela furnace*, holding a batch of platinum at the metal's "freezing" temperature of 1769.3 °C, is designed to resemble such a radiating box as close as possible.

The Bridge from Light to Energy

Past definitions of the candle power and the candela resembled prescriptions for the making of a standard light source rather than links to the rest of the units of measure. Although such "stand-alone units" in a given field shouldn't be ruled out entirely, a better approach is to define the intensity of light by the energy it radiates and thus keep it linked with the units in the International System of measures. Way back, this idea popped up in the definition of the venerable spermaceti candle in the clause "consuming 120 grains of wax per hour." Since wax is the candle's fuel and

source of energy, the quantity consumed is a measure of the radiating energy emitted by the flame. However, not all the energy stemming from a flame, or for that matter, any other light source, can be seen. The graph in Figure 8.3 shows the energy output from a radiating black body at the temperatures of 1000 °C, 1500 °C, and 2000 °C, along with the curve for solidifying platinum (1769.3°) in the candle-power crucible: most of the energy is emitted as invisible infrared radiation, while the portion inside the band

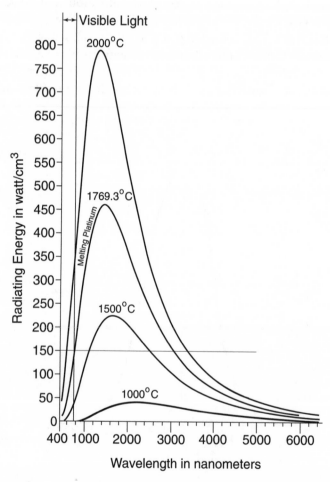

Figure 8.3 Radiating energy and wavelength: energy per second and per centimeter of spectral bandwidth

of visible wavelengths falls sharply from red to cyan. Only the peak energy of radiation from an object as hot as the sun's surface (5714 °C) is located in the visible part of the spectrum.

Variable Human Sight

Fundamentally, luminous intensity is the expression for the degree of response of the human eye to incident light. This makes the sensitivity of human vision a prime factor in the definition of such units.

Figure 8.4 shows the eye's sensitivity at various wavelengths. We perceive the light's wavelength as its color. Blue is linked to shortest waves, red to the longest, and green, yellow, and orange in between. Our eyesight's highest sensitivity lies in the green part of the spectrum, at a wavelength of 555 nanometers (nm). This is also the sunlight's wavelength of highest intensity, which sug-

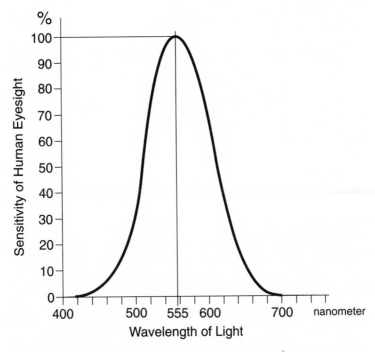

Figure 8.4 Sensitivity of the human eye at different wavelengths

gests a human sense organ, the eye, evolved in response to environmental conditions.

The graph also shows the thresholds of light perception at 400 nm and 750 nm, respectively. Below, we see the ultraviolet band of radiation, and above it, the infrared, also known as *radiant heat.* Switch an electric plate on and you can feel the heat by placing your hand above it (i.e., by detecting infrared) long before the plate heats up and glows red. The response of human vision to light at the wavelengths of 510 nm and 610 nm, respectively, is a mere 50% of that at the "preferential" wavelength of 555 nm. What we see, therefore, is only a smudge of how much light is emitted. Instruments have the same detection problem. A massive radio telescope is not equally sensitive to all wavelengths of radio waves emitted by a cloud of interstellar gas. Scientists always have to remember to remove the effects of the detector from their data.

The interdependence of radiating energy and wavelength on one side, and the human eye's sensitivity and wavelength on the other, limits the usefulness of a setup like the Bunsen photometer (Fig. 8.2) to cases where the light sources on either side of the screen are of the same type and temperature.

The Energy-Based Unit of Luminous Intensity

A source-independent unit for luminous intensity has to be defined at a unique wavelength of luminous radiation. There again, the wavelength of the human eye's maximum sensitivity, 555 nm (5550 angstrom), or 540×10^{12} Hz of frequency, is the obvious choice. The unit of luminous flux, the lumen (lm), has thus been defined as 1/683 watt of luminous radiation at this particular wavelength, channeled through the spatial angle of one steradian. For the ease of interpretation, think of the steradian as the area of one square meter seen from one meter of distance, or, if you like, of one square foot from one foot away.

With 4π for the surface area of a sphere of radius $R = 1$, 1/683 watt per steradian amounts to $4\pi/683 \approx 1/54$ watt of total radiation. If we attempted to build a unit candle to meet these new specifications, it would have to emit 1/54 watt of green light at the wavelength of 555 nm. But such an offspring of the classic

candela would be difficult to build and has been purged from the International System of Units.

Like its predecessors, the magnitude of the lumen has been defined with the aim of keeping the tables of photometric quantities virtually unchanged.

Note that the lumen-related wattage refers to the proper energy of the light beam. The energy it takes to generate such a beam is always much greater. Incandescent lightbulbs are not really designed to generate light, but rather to heat an electric resistor, the filament. Light pops up merely as a kind of by-product. A measure of how good a light source is can be the amount of lumens it provides relative to the amount of watts you supply to get those lumens. Old-fashioned carbon filament bulbs emit 3 to 5 lumen of light per watt of energy input. Modern incandescent lightbulbs yield 10–25 lm/W, mercury lamps 34–40 lm/W, sodium lamps 40–65 lm/W, fluorescent tubes 28–47 lm/W, and Xenon high pressure bulbs 25–30 lm/W. We are mere amateurs, compared to nature: the firefly and some deep-sea marine life produce light by biochemical reactions with an efficiency of up to 80%. Obviously, we have a long way to go until our street lamps can compete with the firefly.

It's Not as We See It

The International System uses cd/m^2 (candela per square meter), for *luminance,* the modernized term for *brightness.* This unit is the offspring of the absolute system's *stilb* (from Greek *stilbein* = to shine), which was the equivalent to one candela per square centimeter. The candela per square meter is thus equivalent to 10^4 stilb. Looking into a light source of above 20 stilb can cause permanent damage to the retina. Beware of looking into the Sun, whose surface's luminance lies between 100,000 and 150,000 stilb. Ptolemy, author of the *Almagest,* went blind during his quest to follow the day-to-day changes of sunspots. Mother Nature provided us with an incredibly wide range of visual perception, starting at luminous stimulations as minute as 1.3×10^{-17} watt-seconds.

The wide range of the human eye's sensitivity is made possible because our perception of light intensity depends on the *log-*

arithm of the energy of the incoming light beam. This so-called *psychophysical law of luminous intensities* is best illustrated by the scale of stellar magnitudes, whose roots stretch back to the Hellenistic philosopher Hipparchos. In A.D. 120, he grouped the visible stars into six classes of magnitude. The fifteen brightest stars of the northern celestial hemisphere became first class, while the sixth class covered the faintest stars that one can still see.

Later, when precise photometers—instruments that objectively measure luminous intensity—were used for measuring starlight, each class of brightness became $\sqrt[5]{100}$ = 2.512 times fainter than the preceding. A first-magnitude star thus shines 2.512 times brighter than one of second magnitude, which itself is 2.512 times brighter than a star of third magnitude, and so on. Nevertheless, physiologically we perceive geometric progression as equal steps from six to one. Knowing stars' magnitudes can help astronomers—both amateur and professional—observe transient phenomena. Meteor showers and a firebolt that streaks across the sky for a brief few moments before exploding can have their maximum brightness assessed—thanks to Hipparchos.

On this logarithmic scale, Venus, the evening and morning star, measures –4.0 at its brightest; the full Moon –12.6; and the Sun –26.9. Air pollution has cut down on the visibility of the fainter stars, so that only stars of third magnitude or brighter can be seen in urban and suburban areas. One has to seek remote areas in a dry climate to glimpse the five thousand stars on our celestial hemisphere, and the world's mightiest observatories are placed on the highest possible mountain tops. The day is approaching when only the price of a ticket up to the space station could buy us that kind of view.

And It's Not as We Hear It

A musical event, announced at the time as "the biggest concert of all time in the biggest hall in the world, with the largest orchestra and largest choir," took place in 1872 at Boston's St. James Park. With the help of a hundred assistant conductors and orchestras, the famous Viennese composer and concert master Johann Strauss directed an orchestra of 20,000 performers as they let

loose the sound of "The Blue Danube Waltz" for an audience that numbered around a hundred thousand.

Did the sound energy of that "greatest ever" band of musicians —one hundred times that of a routine rendering of the master's melodies—roll like thunder over the city, shaking its very foundations, just as the walls of Jericho were destroyed by the sound of the shofar? Obviously not. The spectators must have been quite happy with the volume of sound they got. Otherwise, they wouldn't have paid so handsomely for the most cherished souvenir of the event, a lock of hair from the maestro's ebony coiffure. It later turned out that the locks came more often than not from his equally black Newfoundland dog, which Stephen, the maestro's enterprising valet, realized was a more prolific lock supplier than his master.

But why did such a massive accumulation of musical instruments not produce a sound loud enough to burst the eardrums of the members of the audience? What made the sound from such a monster orchestra bearable is that the psychophysical law of stimulation and perception applies to sound in much the same ways as it does for light. In acoustics, it lead to the well-known unit of bel for loudness, because logarithmic perception is characteristic of our auditory system as well.

Expressed in bel (1 bel = 10 decibels or dB), the loudness ratio of two sound sources with intensity levels of I_1 and I_2 becomes $\log(I_2/I_1)$. The threshold of audibility, the noise you get when a falling leave lands on a groundcover of dried leaves, lies at zero bel loudness. Starting from there, the rustle of leaves in a breeze reaches the 20 dB (2 bel) point, speech 60 dB, and thunder and jet engines 120 dB and above.

If, in the Johann Strauss story, I_R stands for the intensity of sound from each of his one hundred orchestras, a band of one hundred such assemblies would bring this up to $100 \times I_R$. The loudness ratio is $\Delta L = \log(100 I_R / I_R) = \log 100 = 2$.

Thus, Bostonians heard the beautiful "Blue Danube" waltz from one hundred orchestras a mere 2 bels (20 dB) louder than what a single band of musicians would have managed. Likewise, the sound from a band of ten rock singers is merely one bel louder than one lone singer's voice. Heavy metal could never have existed if nature hadn't provided our auditory faculties with such a safety net.

As with human vision, the logarithmic response of our auditive system leads to a stunning range of audibility. From the lowest perceptible sound levels to the point where hearing becomes painful (1 watt/m^2 or 2×10^4 atm), the range of sonar intensity is $1:10^{12}$.

But physiology is not the only field for logarithmic scales. The Richter scale for the intensity of earthquakes, devised by American geologist Charles W. Richter (1900–1985), specifies eight classes of earthquake intensities, with each consecutive number standing for a tenfold increase in magnitude.

But It Is Pretty Much as It Feels

Only tactile sense needs higher levels of stimulation. The lower threshold lies in the range of 0.2 to 0.4×10^{-6} joules, which is probably about the impact energy of a mosquito aiming its proboscis at some tender region of our epidermis. Yet the range of tactile perception is still amazing, at least if our upper threshold is the feel of a punch from Max Schmeling or Mohammed Ali.

· 9 ·

Hot Stuff

Temperature, Pressure, and Thermodynamics

About 400,000 years ago, *Sinanthropus pekinensis*—the so-called Peking man—developed the ability to sustain a fire, which must have been started naturally, perhaps by lightning. Excavations of sites from the mid- and young Paleolithic epoch provide the first evidence of humans' ability to generate fire for themselves, using flint and tinder. The fire quirl, essentially a wooden spindle rotated by one loop of a bow's cord slung around it, produced heat by friction as it was seesawed on a kind of needle bearing. Because a quirl is made of perishable materials, they have not been found intact. Archeologists piece together the fragments of wood or leather that remain and infer what the quirl must have looked like and how it produced fire. Such evidence, while compelling, is not incontrovertible.

From the times of the Roman Empire to the seventeenth century, flint stones coated with sulfur were popular as fire starters still, fire preservation played an important role in the early Middle Ages. At nightfall, the "curfew"—a fire cover made of perforated brass sheet—was set over a heap of hot coals taken from the fireplace, which usually preserved the glow until the following morning. In 1068, under the reign of William the Conqueror, the bells were rung at 7 P.M. to remind people to cover the fire and extinguish the lights. This was a two-pronged law, for it combined the

control of fire hazards with the prohibition of nocturnal assemblies. William might have been more concerned with the latter than the former. He had ascended to the English throne when his Norman troops defeated the Saxon king, Harold Godwinson, at the Battle of Hastings some two years earlier. No doubt he slept more soundly knowing the Saxon earls couldn't plot his overthrow in secret nighttime meetings.

Beyond Europe, a fire-pump, a fascinating device, was used in Indo-China and Indonesia. It consisted of a wooden cylinder with a wooden rod sliding inside like a piston. Repeatedly lowering the piston quickly generated heat by compressing the air in the cylinder until a pinch of tinder, housed in a groove at the lower end of the piston, began to glow.

Throughout history, fire has been the means by which metals, such as gold, silver, copper, lead, tin, and zinc, were melted from their ores. To reach the high temperatures needed for iron making, forced ventilation was usually provided by a pair of bellows. But the concept of heat came only in the eighteenth century with Georg Ernst Stahl and Johann Joachim Becher's "phlogiston" theory. Similar to the "world ether," phlogiston—from the ancient Greek verb *phlogizein* meaning to set on fire—was conceived of as an infinitely tenuous form of matter that flowed from the hot regions of a body into the cooler ones, and played an important role in chemical reactions. Although plagued by inherent contradictions, the search for proof of phlogiston's existence led to the discovery of a great number of metals and of most naturally occurring gases. It was James Watt's invention of the steam engine that sounded the death knell for phlogiston and lead to the interpretation of heat as thermal energy.

Heat is a form of energy and it is the kinetic energy of the molecules and atoms that compose matter. What we perceive as an object's temperature merely reflects the intensity of its atomic motions, which even determine the state of matter. Ice, for instance, consists of water molecules (H_2O) oscillating within the confines of a hexagonal crystalline lattice. Above the freezing point the molecules break their structural bond and change phase from ice to liquid water. At the boiling point, water molecules break free from one another, because the energy of their motion

typically exceeds cohesive energy that binds them together: the state of water shifts to that of a gas we call steam. Steam is transparent and thus invisible, and what we see pouring from a kettle is actually condensed water formed when the steam molecules have cooled off by contact with the surrounding air.

Lowering a body's temperature amounts to slowing down the oscillations of its atoms or molecules, which at −273.15 °C (absolute zero) come to a virtual standstill. Still lower temperatures are therefore impossible to achieve.

The Mercury Thermometer

Internal vibrations can also cause matter to expand with rising temperature. Water is an oddity, because it expands upon freezing, which is why water pipes can burst in cold weather. Mercury, a metal in liquid state at room temperature, expands by 0.000181 (0.0181%) of its volume per degree Celsius (1.8 °F) and therefore 1.81% from 0 °C (freezing point of water) to 100 °C (boiling point of water). The classical mercury thermometer converts this minute expansion into a readable column by channeling the mercury that fills a bulb into a very thin, flat capillary tube.

For years, parents had to struggle with thermometers made of glass, which shatters, and filled with mercury, which is poisonous. Nowadays, new types, such as the Tympanic thermometer, can be inserted into a small child's ear and "clicked" to find the temperature. They use the thermoelectric effect for precision temperature readings. A pair of thirty-six-gauge copper-constantan wires are soldered at the tip and drawn into fine polyethylene tubing. A temperature difference between the two ends of the wires causes an electric current to flow, which is turned into a readout on a liquid-crystal display. No shattering glass, no poisonous liquids, and no need to wait several minutes for the mercury to rise in the tube. It's much better for all concerned!

Rudimentary thermometric devices date back to the early seventeenth century. But in the year 1714, a doctor by the name of Daniel G. Fahrenheit from the city of Danzig (now Gdansk) in Poland built calibrated mercury and alcohol thermometers. Originally, Fahrenheit started his scale with the zero degree mark at

the freezing temperature of an equal parts ice-salt mixture. He placed the freezing point of water at 30° and human body temperature at 90°, but later revised those values for a closer match with the medical thermometers of those days, to 32° and 96°, respectively. On the resulting scale, freezing and boiling points of water fell 180° apart, at 32° and 212°. But Fahrenheit was still somewhat off the mark, for in the US, average human body temperature is 98.7 °F; in the U.K., the cold-blooded Brits average only 98.4 °F.

Freezing and boiling temperatures are always given for the sea level barometric pressure of 760 mm mercury. Water boils sooner at lower pressure. For instance, in the thinner air at 1900 meters of altitude, the boiling point is down at 200 °F. That makes pressure cookers a must for tenderizing meat in mountain restaurants, and it also means it is impossible for mountain climbers to have a hot drink on Mount Everest. On the other hand, the water's freezing temperature is almost immune to pressure changes: 317 times normal atmospheric pressure (317 atm = 4650 psi) lowers it by a mere 5 °F.

The Celsius Scale

Though we remember Anders Celsius mostly as the originator of the centesimal temperature scale, his role as a scientist is much larger than that. Known in his homeland as the "founder of Swedish astronomy," he purged Uppsala University of antiquated Ptolemaic concepts and its domination by theologians. Even on his deathbed, Celsius's critical thinking became evident when a hastily summoned clergyman went overboard in his efforts to sweeten his parishioner's last hours in this world, providing glowing descriptions of the world beyond. Celsius soberly reminded the clergyman he would soon enter a stage where he could personally check out every word he was hearing.

Most of us are familiar with the centesimal temperature scale, with its 100° interval between freezing and boiling points of water, which Anders Celsius proposed about three decades after the introduction of the Fahrenheit scale. However, his original thermometer counted backwards, letting water freeze at 100 °C

and boil at 0 °C. Luckily, the famous Swedish physician and botanist Carl von Linné put things into their logical order and thus made himself a co-sponsor of the modern Celsius temperature scale. Linné, known to his contemporaries as Carolus Linnaeus, is often called the Father of Systematic Botany due to his scientific system of classifying plants and animals: he gave each specimen two Latin names, the first for the genus, the second for the species.

Absolute Temperature

A century later, Sir William Thomson, a.k.a. Lord Kelvin of Largs, proposed an absolute temperature scale. Although refrigeration had not yet been invented, Kelvin deduced a pretty precise value for the lowest physically possible temperature. He inferred this from his observation that for every 1 °C drop in temperature, a gas will contract almost uniformly by 1/273 of its volume at 0 °C. So, thought Kelvin, there would be a volume of zero at –273 °C. Matter cannot physically disappear, and yet, Kelvin's extrapolation came surprisingly close to the correct value of –273.15°C for the absolute zero point.

The absolute temperature scale, fittingly named the Kelvin scale, thus starts with zero degrees kelvin at –273.15 °C. It maintains the division of the water's freeze-to-boil temperature range in 100 degrees, so that the water's freezing and boiling points are located at 273.15 and 373.15 degrees kelvin, respectively.

Cryogenics

But how to get down to rock-bottom temperatures? In 1895, Carl von Linde invented a process for liquefaction of air through successive compression and decompression, interspersed with refrigeration by precooled air in a countercurrent.

This same process, yet with the use of liquefied air (–193 °C) as the coolant, allowed for the liquefaction of gases of still lower boiling temperature, such as nitrogen (–196 °C). This, in turn, was used as a cooling fluid in the liquefaction of hydrogen (–252.8 °C), and ultimately of the rare gas helium (–268.9 °C). Allowing

part of the liquefied helium to evaporate makes the remaining batch still cooler. If this is done within a strong magnetic field, subsequent removal of the field causes a further drop in temperature. Known as *adiabatic demagnetization,* this process brings temperatures down to within a few ten-millionths of a degree of the absolute zero point. However, reaching this point itself is impossible, because it would take infinite energy.

At such incredibly low temperatures, new physics can be found. A supercooled banana can be shattered with a hammer. Superfluids can flow without any viscosity. Groups of atoms can join together and behave in entirely new ways in a state known as a Bose-Einstein condensate, named for Albert Einstein and the Indian physicist S. N. Bose. Physicists hope to harness all of these properties to create new devices of practical use at everyday temperatures.

Temperature-Scale Conversion

The Kelvin scale became standard in the International System of Weights and Measures as the *Thermodynamic Temperature Scale,* often called the *Absolute Temperature Scale.* The word "degree" and the symbol ° are omitted from the abbreviation K, so that, for instance, the correct notation for the freezing point of water is simply 273.15 K.

Kelvin and Celsius degrees are of identical magnitude, but while the zero point of the Celsius scale is the freezing temperature of water (0 °C = 32 °C), the Kelvin scale begins at absolute zero (−273.15 °C). The following formulas show the conversion among temperatures in Kelvin, Celsius, and Fahrenheit:

$$\text{Celsius to Kelvin:} \qquad K = {}^\circ C + 273.15 \qquad (9.1)$$

$$\text{Fahrenheit to Celsius:} \quad {}^\circ C = \frac{5}{9} \times ({}^\circ F - 32) \qquad (9.2)$$

$$\text{Celsius to Fahrenheit:} \quad {}^\circ F = \frac{9}{5} \times {}^\circ C + 32 \qquad (9.3)$$

Not surprisingly, our customary FPS system created its own absolute (Rankine) temperature scale by extrapolating the Fahrenheit scale

down to the absolute zero point. Combining Eqs. (9.1) and (9.3) places the zero point of the Rankine scale at $32 - 273.15 \times \frac{9}{5} =$ -459.67 °F. Rankine temperatures can thus be figured by adding $459.67°$ to readings in the Fahrenheit scale, that is:

$$\text{Fahrenheit to Rankine:} \quad °R = °F + 459.67. \quad (9.4)$$

The mutual relations of these temperature scales and the fixed points commonly used for their calibration are graphically shown in Figure 9.1.

A Pressing Matter . . .

As with temperature, pressure helps determine the state of matter. But unlike the basic unit of temperature, pressure, the force per unit area, is a derived (compound) unit.

The pressure of the water from a tap at the foot of a water tower is caused by the weight of the water in the pipe that leads down from the reservoir. We may express it simply in meters (or feet) of "water head" as the height difference between the spud from where the water is drawn and the liquid surface in the reservoir.

In a pipe of one-square-centimeter internal cross section, each centimeter of height contains one cubic centimeter of water and herewith weighs one gram. Thus, a water column 10 meters high weighs $10 \times 100 = 1000$ g = 1 kg and exerts a pressure of 1 kg/cm^2 on its base, i.e., at the tap. This unit, the equivalent of 10 meter of water head, is the *technical atmosphere* (at) in the Technical System of Measures.

By coincidence, atmospheric pressure equals the technical atmosphere whenever the barometer stands at 735.6 mm. That's only a tad less than the 760 mm meteorological standard at sea level (zero altitude) and 0 °C, which has become the unit of *physical atmosphere* (atm).

Altogether, we have the conversion 1atm = 1.033227 at = 1.033227 kg/cm^2 = 14.696 psi (pounds per square inch), so that the usual 30 psi pressure in a car's tire is slightly over 2 at. But let's keep in mind that we count physical atomspheres from absolute vacuum on, while all the other units stand for pressure above ambient pressure.

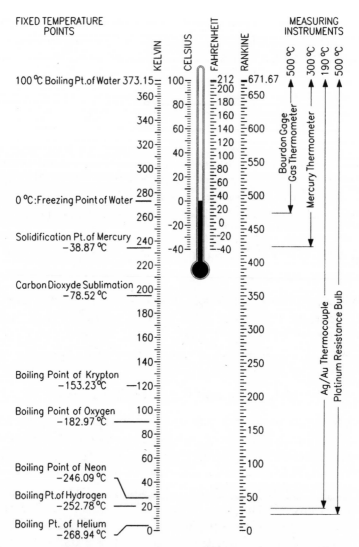

Figure 9.1 Thermometric scales, their fixed gauging points, and temperature measuring methods

The International System's (SI) unit for pressure, the *pascal* (Pa), is deduced from the newton (N), the basic unit of force, and the meter (m), as one newton per square meter, 1 N/m^2. Thus, 1 Pa is 1 N/m^2 = 0.1019716 kgf/m^2 ≈ 10^{-5} kgf/cm^2. Since 1 kgf/cm^2 equals 10 m of water head or 10 × 1000 = 10^4 mm,

we see that one pascal is about $10^{-5} \times 10^4 = 0.1$ mm of water head. The more substantial unit of kilopascal (kPa) comes to approximately $1000 \times 0.1 = 100$ mm of water head.

The US Equivalents

In our customary FPS system, the pound per square foot would be the pascal's counterpart, but no such unit has been made official. Thus, the classical pound per square inch is still going strong.

The metric conversion of the psi follows from the National Institute of Standards and Technology (NIST) relations of 2.54 cm to the inch, and 0.45359237 kilogram·force to the pound, as

$$1 \text{ psi} = \frac{0.45359237}{2.54^2} = 0.070307 \text{ kgf/cm}^2 = 703.07 \text{ kgf/m}^2.$$

This equals $703.07 \times 9.80665 = 6894.76$ N/m^2 or 6894.76 Pa. Inversely, the kilopascal (1 kPa = 1000 Pa) equals 0.145 psi, and the megapascal (1 MPa = 10^6 Pa) 145.04 psi. That makes a car's tire pressure of, say 30 psi, into $30 \times 6894.76 = 206753$ Pa ≈ 207 kPa.

Absolute and Relative Pressure

While the technical atmosphere (at) and our pounds per square inch (psi) stand for pressure above ambient air pressure, physical atmospheres (atm) express absolute pressure. "Zero atm" is thus the equivalent of total vacuum, while "zero at" means ambient atmospheric pressure. Adding 1.033 kfg/cm^2 converts relative pressure into absolute pressure, much like the addition of 273.15° converts degrees Celsius to degrees Kelvin. For example, 10 at (10 kg/cm^2) of relative pressure equal 10 + 1.033 = 11.033 kg/cm^2 of absolute pressure, that is 11.033/1.033 = 10.69 atm. Likewise, 10 physical atmospheres of absolute pressure equal $(10 - 1) \times 1.033 = 9.30$ at, that is, 9.30 kg/cm^2 of relative pressure.

In Figure 9.2, a mercury U-tube pressure gauge shows the cabin pressure change from relative to absolute when a rocket leaves the atmosphere. The gauge reads zero pressure on the

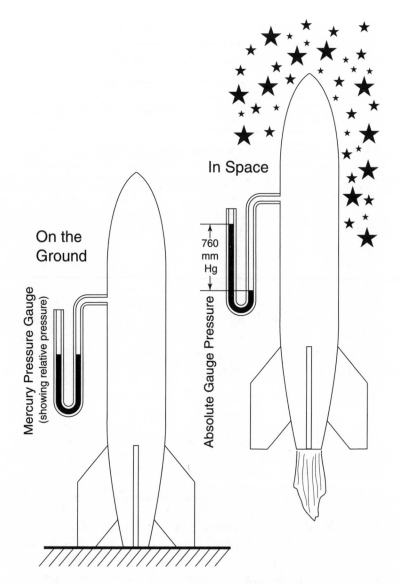

Figure 9.2 The U-tube barometer shows relative cabin pressure with the rocket on the ground and absolute cabin pressure with the rocket in space.

ground, but 1 atm (760 mm Hg) in the vacuum of space, though the pressure inside the cabin remains constant.

A Matter of Convenience

Many problems of thermodynamics can be solved by simple logic if we use the atm as our unit of choice. For instance, it is obvious that a tank, sized to hold, say, 1 kg of a certain gas under the ambient pressure of 1 atm, could contain 2 kg of gas when pressurized to 2 atm, and 3 kg at 3 atm, etc.

Tire pressure of, say, 30 psi, converts to $30/14.696 + 1 = 3.042$ atm. This shows, without any further preliminaries, that we need about three times the tire's internal volume of compressed air to pump it up.

Underwater pressure follows from the relation of 10 m of water head per technical atmosphere ($1 \ kg/cm^2$). For instance, robotic submarines used to explore the wreck of the *Titanic*, some 12,600 ft down on the ocean floor, must withstand $12,600 \times 0.3048 = 3840$ m of water head, or $3840/10 = 384$ at. Such problems are equally solvable in pascal units, but at the cost of considerably more complex arithmetic.

The Mercury Barometer

The practice of expressing air pressure by the height of a mercury column (mm Hg) stems from the traditional instrument for measuring atmospheric pressure, the mercury barometer. Of far better precision than most commercial instruments, the mercury barometer was developed in 1643, when Evangelista Torricelli filled a closed-end glass tube, about 3 feet long, with mercury. Turned down into a dish of mercury, only a small part of the filling flew out. A 30-inch column of mercury stayed in the tube because ambient air pressure held it up. While Torricelli's interest focused on the vacuum under the closed end of the tube, he had unwittingly created the first barometer, which later allowed him to discover periodic changes in atmospheric pressure.

Updated models (Fig. 9.3) function on the same principle but use a U-tube mounted on a vertically adjustable bracket. Thus,

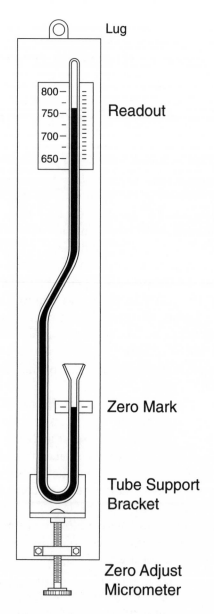

Figure 9.3 The classical U-tube mercury barometer

the lower end of the mercury column must be aligned with the zero mark of the instrument's fixed scale before each reading.

Seasonal changes aside, atmospheric pressure at sea level has been standardized as 760 mm = 29.92 inches of mercury head.

The Sipping Sound of Vacuum . . . an Illusion

When we sip our soda through a straw, we lower the pressure in our mouth. Then the atmospheric air pressure on the liquid surface in the cup propels the refreshment up the straw. Likewise, atmospheric pressure at the open end of a mercury barometer's U-tube, rather than the vacuum in the tube's upper portion, keeps the mercury column from dropping down.

Although the scales of some industrial vacuum gauges list negative values, nature's zero pressure is the absolute vacuum of outer space, so that any gas pressure, minute as it may be, is physically positive. For this reason, a shallow garden well must not exceed 23 feet deep, for no commercial pump can "suck" water up from farther down than that. The reason is that a pump does not suck water up. Atmospheric pressure forces the groundwater up into the partially evacuated admission side of the pump, from where the rotor swirls it to our lawn sprinklers. And since the maximum pressure available for lifting the water up is ambient atmospheric air pressure of 1.033 kg/cm^2 at best, corresponding to $1.033 \times 10 = 10.33$ m of water head, this sets the theoretical limit for the depth of a well unless the pump is installed farther down. As commercial water pumps rarely generate anything better than a 70% vacuum, seven meters of lift is what one may realistically expect.

Likewise, the domestic vacuum cleaner does not suck dust out of the carpet. Rather, the gadget's blower creates a partial vacuum at its intake, causing atmospheric pressure to force air into the blower and carry the dust with it.

Negative pressure annotations are a human-made format for expressing pressure levels below atmospheric pressure (29.92 inches or 760 mm of mercury head at sea level). Readouts from vacuum gauges, such as "minus 29 inches" or "29 inches of vacuum," must thus be interpreted as ambient barometric pressure minus 29 inches, or $29.92 - 29 = 0.92$ inch (of mercury column) of absolute positive pressure.

The Not-So-Perfect Gas

Although a perfect gas doesn't exist, many gases, such as air, come pretty close. If we slowly crank a car engine with a, say, 1:8 compression ratio, air in the cylinders gets compressed to 1/8 of its original volume. Thus, we intuitively expect internal cylinder pressure to rise from 1 atm at admission to 8 atm at the end of the compression stroke. This relation of volume and pressure for a perfect gas was first formulated in 1662 by the chemist Robert Boyle (1627–1691) and named Boyle's law:

$$p \times V = \text{constant} = \mathfrak{R}_1, \tag{9.5}$$

where p stands for the absolute pressure, and V for the volume of a given amount of gas.

For the dependence of volume and temperature, Joseph Louis Gay-Lussac found a similar relation almost a century later, namely the proportionality, at constant pressure, of volume and absolute temperature T, expressed by the formula

$$\frac{V}{T} = \text{constant} = \mathfrak{R}_2. \tag{9.6}$$

These two equations can be combined into one form of the "gas equation":

$$\frac{p \times V}{T} = \text{constant} = \mathfrak{R}. \tag{9.7}$$

Thus, the volume V of a given amount of a gas is proportional to its Kelvin temperature T, and inversely proportional to gas pressure p.

Applied to one cubic meter of gas under the so-called Standard Conditions of Thermodynamics, namely $p = 1$ atm of pressure and $T = 0\ °C = 273.15$ K temperature, the proportionality constant $\mathfrak{R}$ follows from Eq. (9.7) as

$$\mathfrak{R} = \frac{1 \times 1}{273.15}.$$

Introduction of the symbol T_0 = 273.15 makes the constant $\Re$ = $1/T_0$ and brings the gas equation into the elegant form

$$p \times V = \frac{T}{T_0} \text{ or } V = \frac{T}{p \times T_0} \text{ or } \frac{p \times V}{T} = \frac{1}{T_0}. \quad (9.8)$$

From this, we deduce the volume of the cubic meter of gas under the standard conditions of p = 1 atm = 760 mm Hg, T = 0 °C = 273.15 K, as

$$V = \frac{T}{p \times T_0} = \frac{273.15}{1 \times 273.15} = 1 \text{ Nm}^3$$

This unit volume of any gas, called the *Normal cubic meter*, weighs as much as the weight density of the gas, for instance,

1 Nm^3 of hydrogen = 0.08988 kg,

1 Nm^3 of oxygen = 1.4290 kg,

1 Nm^3 of air = 1.293 kg.

Applications of the Gas Equation

For two different states of a gas, marked by the subscripts 1 and 2, the gas equation becomes

$$\frac{p_1 \times V_1}{T_1} = \frac{p_2 \times V_2}{T_2}. \quad (9.9)$$

This equation can save us money the next time we buy bottled gas. For instance, oxygen for gas welding is stored in seamless steel cylinders of 1.625 cubic feet (cft) capacity under 2200 psi of pressure (Fig. 9.4). To find how much oxygen we really get, we initially convert psi to physical atmospheres by use of the relation p_1 = 2200/14.696 = 150 atm. p_2 is the atmospheric pressure of 1 atm. Further, we figure from the relation of 0.3048 meter to the foot that one cubic foot equals 0.3048^3 = 0.0283 m^3, so that the cylinder contains 1.625 × 0.0283 = 0.0460 m^3. For the ease of convertibility into weight, we want the result in Nm^3, which we get

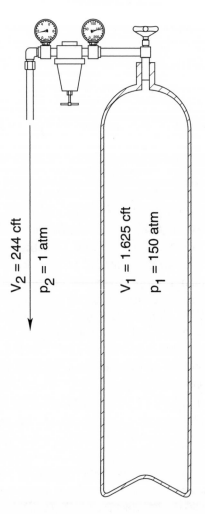

$V_2 = 244$ cft
$p_2 = 1$ atm

$V_1 = 1.625$ cft
$p_1 = 150$ atm

Figure 9.4 Oxygen cylinder

by setting $T_2 = 0$ °C. T_1 is the ambient temperature of 20 °C. Thus, we get from Eq. (9.9):

$$\frac{150 \times 0.046}{293} = \frac{1 \times V_2}{273} \text{ and } V_2 = \frac{150 \times 0.046 \times 273}{293}$$

$$= 6.43 \text{ Nm}^3.$$

Since 1 Nm^3 of oxygen weighs 1.429 kg, that equals 6.43 × 1.429 = 9.2 kg, the amount that should be invoiced for the refill.

A Crazed Thermometer

Rearranging the terms of the gas equation Eq. (9.7) into

$$T = \frac{1}{\Re} \times p \times V$$

would make it resemble Newton's second law, $F = m \times a$, if it weren't for the proportionality constant $1/\Re$.

In the case of Newton's second law, the introduction of the unit "newton" for force made the proportionality constant equal 1. No similar attempt has ever been tried for the gas equation, but here, too, it could be done by setting $\Re = 1/T_0 = 1$, which makes $T_0 = 1$. This would give rise to a new temperature scale, which we'll call °H, yet leave all the laws of physics unchanged.

Such a scale would start at absolute zero temperature and place the freezing point of water at $T_0 = 1°$. The boiling point then locates at 373.15/273.15 = 1.366°H. For the sake of convenience, one would use the subunit "centesimal degree" (0.01°) in order to extend the water's freeze-to-boil span to 36.6°, reaching from 100° to 136.6 °H.

Of course, we wouldn't want such a scale as part of our daily life, the more so since the gas equation is based on a merely theoretical substance, the "perfect (or ideal) gas," which has no exact counterpart in nature. For a closer approximation to the behavior of real gases, the Dutch physicist Johannes Diderik van der Waals introduced the refined gas equation that carries his name:

$$\left(p + \frac{a}{V^2}\right) \times (V - b) = R \times T.$$

The constants a and b, both in J · m^3/mole for the most common gases are as shown in the following:

Type of Gas	a	b
Hydrogen	0.0247	2.65×10^{-5}
Helium	0.00341	2.34×10^{-5}
Nitrogen	0.1361	3.85×10^{-5}
Air	0.1358	3.64×10^{-5}
Water H_2O	0.5507	3.04×10^{-5}
Ammonia NH_3	0.4233	3.73×10^{-5}
Carbon Dioxide CO_2	0.3643	4.27×10^{-5}
Freon CCl_2F_2	1.078	9.98×10^{-5}

While the original version of the kinetic gas theory, which leads to a deduction of the gas equation, operates with pointlike gas molecules that bounce elastically against the walls of the container and against one another, van der Waals's equation follows if we consider molecules of a certain size. Here, the molecules can't get too close together, generating a "pressure" that is modeled with the a/V^2 term. The $(V - b)$ term comes from the fact that a molecule within the volume V of the container can move only in the space not occupied by other molecules.

The van der Waals equation is more complex than the perfect gas equation, and in it the parameters a and b have to be determined empirically, which no doubt accounts for its less frequent use.

The Two Faces of the Gas Equation

The traditional form of the gas equation

$$p \times V = n \times R \times T, \qquad (9.10)$$

with the gas constant R as the proportionality constant, is based on the amount of n moles of the gas under investigation, while Eq. (9.8) applies to the normal cubic meter (Nm^3) for the quantity of gas. Therefore, do not mistake the constant $\Re$ in Eq. (9.7) et seq. for the universal gas constant R.

One mol of a given gas of molecular mass M embodies M gram, and one kilomole (kmol) M kilogram of that gas. One kmol of hydrogen weighs 2 kg, and one kmol of oxygen weighs 32 kg.

In theory, that relation should make the mole of all gases equal in volume, typically 22.414 m³. Nevertheless, minor differences exist.

Where Units Become a Must

In 1977, two geologists descended 8200 feet (2500 m) into the ocean near the Galapagos Islands in hopes of discovering deep-sea hot springs or hydrothermal vents. They found the first vent, which, according to John Edmond, a Massachusetts Institute of Technology geologist, covered an area of about 100 meters across, where water at 60 °F "streamed out of every orifice and crack in the sea floor." If a bubble of methane gas emerged from the vent and reached the surface, it would have gone from 250 at of water pressure at the sea bottom to about 1 atm near the surface. According to Eq. (9.5), rewritten into $V_1/V_2 = p_2/p_1$, its volume would increase 250/1.033 = 242-fold.

Typically, this kind of relation between two variables, like the volume and pressure of a quantity of gas, are readily expressed by a set of dimensionless equations. However, introduction of units makes it possible to include a third pertinent variable, for instance temperature, as in the gas equation (equation of state). The price we pay for the convenience of joining three variables in one single formula is that it works only within a properly structured system, in particular the SI, where p comes in pascal, V in cubic meters, and T in kelvin. Beware of using that same formula with foot and pound units unless you previously converted the gas constant R into units of ft·lbf/lb·mol·°R.

Part of the energy it takes to compress a given volume of gas shows up, in the end, as heat. If that heat is given time to dissipate, the process is *isothermal*, and Boyle's law applies. On the other hand, the compression stroke of an internal combustion engine (Fig. 9.5) happens so fast that there is no time left for any significant loss of thermal energy through heat exchange between the gaseous charge and the cylinder walls. The process becomes *adiabatic*, and the pertinent equations of state are pV^γ = constant. Herein, γ is the quotient of the specific heat capacity of the gas in point at respectively constant pressure and at constant volume. For the air/gas mixture in an internal combustion engine, γ = 1.41. Thus we get

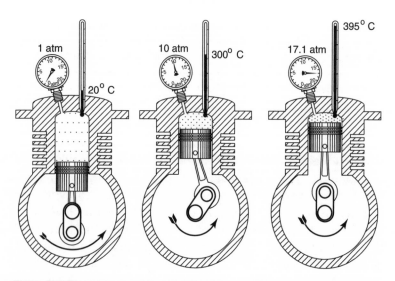

Figure 9.5 The adiabatic compression stroke of an internal combustion engine

$$\frac{p_2}{p_1} = \left(\frac{V_1}{V_2}\right)^{1.41} \text{ and } \frac{T_2}{T_1} = \left(\frac{V_1}{V_2}\right)^{0.41}.$$

Note in Figure 9.5 the simultaneous rise in pressure and temperature which, in Diesel engines, is high enough to cause spontaneous ignition of the jet of pulverized fuel injected a split second prior to the piston's arrival at its uppermost position.

Sometimes it's tough to know exactly what is isothermal and what is adiabatic. Sir Isaac Newton, when seeking to understand sound waves, already knew they were compressions of air, but he thought they were isothermal. They are not; they are adiabatic, so Sir Isaac calculated a wrong expression for the speed of sound.

The Metric Black Sheep that We Adopted

The popular calorie, printed on food labels in order to make us feel guilty whenever we eat anything other than an undressed salad, cabbage, or oat bran, actually is a unit of thermal energy. Humans combine food chemically with oxygen, a process equivalent to combustion, which is where the calorie comes from.

One hundred calories is the thermal energy (heat) required to bring 1 kg of water from freezing to boiling. 180 BTU (British Thermal Units) do the same for 1 pound of water. The calorie is based on Celsius temperature and the kilogram, while the BTU is based on degrees Fahrenheit and the pound. With 1 lb = 0.4536 kg, this leads to the conversion factors 1 BTU = 0.4536 × 100/180 = 0.252 kcal, and 1 kcal = 1/0.252 = 3.967 BTU. For most applications, a 1:4 conversion factor from kcal to BTU is good enough. Whatever our reasons were for using this metric unit for food labeling, the United States imported the wrong one. The International System recognizes only one unit for work and energy, whether mechanical, electrical, or thermal: *the joule* (J).

The mechanical equivalent of heat, an expression for the relation between thermal and mechanical energy, equates 1 kcal with 4184 J or 427 kgf·m. This suggests that, if our body could fully use the "heat of combustion" of the food we eat, every single calorie would suffice for lifting the mass of about half a ton from the floor to our desk top. A 100-calorie diet ice cream would suffice to lift 50 tons to that height.

But unfortunately, the thermal efficiency of converting heat energy to mechanical energy depends on what's known as the *process temperature,* T_p. Our body temperature is about 37 °C = 273 + 37 = 310 K, which is our process temperature. But the average temperature of the environment, T_E, typically 20 °C = 293 K, is not far behind. The thermal efficiency $(T_p - T_E)/T_p$ of human biological processing is $(310 - 293)/310 = 0.055$, an efficiency of about 5.5%. But even that should enable us to lift $0.055 \times 50 = 2.75$ metric tons up to our desk top before we reach for a second helping of ice cream. Why not check theory by experiment? Volunteers welcome.

· 10 ·

The Missing Link

Energy

"It is later than you think," says a how-to book on changing jobs—hopefully for the better. "They may already have seen that shifting glance in your eyes, or the 'help wanted' page on your lap," it continues.

Something similar could be said about America's conversion to the metric system, because metric units have sneaked into peoples' lives without their awareness or consent.

That the units of time—seconds, hours, and days—are identical in both systems goes without saying. But how many of us realize that familiar units such as the watt, volt, ampere, ohm, coulomb, farad, etc. are also from the metric system?

Paying Metric Utility Bills

Utilities bill kilowatt-hours for a month's supply of electricity that drives the electric motors of washer, dryer, ceiling fan, and power tools and generates the thermal energy in our range, water heater, room heater, and, yes, our lightbulbs. What we are paying for is thus the *work* performed by the electricity we used up, not surprisingly revealing that the kilowatt-hour is a unit of work.

Electricity acts as carrier and distributor of the work done by the primary sources of energy, including mechanical, thermal, and nuclear. Mechanical energy comprises hydroelectric power, wind power, and some less attractive alternatives such as power from

ocean waves and the tides. With mechanical energy generation, you get what you pay for, so to speak. For instance, one metric ton or 1000 kg of water flowing from a lake 100 m above into a turbine generates $100 \times 1000 = 100,000$ meter·kilogram (m·kgf) of work. The conversion factor of 9.80665 from joule to meter·kilogram makes that 980,665 J, or 980,665 watt·seconds, and the relations of 1000 watts to the kilowatt, and 3600 seconds to the hour, convert it into kilowatt·hours:

$$3600 \times 980665/1000 = 3,530,394 \text{ kilowatt·hours}$$
$$\approx 3500 \text{ megawatt·hours.}$$

The power of a plant supplying that amount of work, say, every two hours, amounts to $3500/2 = 1750$ megawatts (MW) in line with the output of the Grand Coulee Dam at the eponymous city in the state of Washington. Grand Coulee is the largest concrete dam in North America and the third largest producer of electricity in the world. Brazil generates 90% of its electricity in hydroelectric power plants. Hydroelectricity is, relatively speaking, environmentally friendly and cheap.

The term "wind power" recalls the picturesque windmills that, in days long gone, were standard for crushing and grinding grain in the Netherlands and northern Germany. Their offspring survive in rural areas of the United States, powering water pumps in the backyards of farmhouses. However, the advent of aerodynamically designed wind turbines brought wind-power plants to the point of "making a difference" in the nation's energy policy. For instance, the Buffalo Ridge Windplant with its seventy-three turbines generates 25 megawatts of electricity for the Northern States Power Company. An increase to 425 megawatts is planned for the near future. Although confined at present to geographic areas of strong and principally steady winds, wind power has a bright future in offshore installations.

Not surprisingly, the power of the tides and of oceanic waves can also be harvested, but so far the output of such plants has failed to reward the necessary capital investment. Still, new designs might be forthcoming that would increase efficiency. A

worldwide oil shortage, and subsequent price hike, might also increase the attractiveness of tidal power.

Two centuries ago, horses were predominant among the primary power sources for getting around and for driving all kinds of machinery. The horsepower (HP), as we know, is based on the work a horse is capable of doing. James Watt, who made the steam engine what it is today, found that a horse that powered a mill for grinding corn walked 144 rounds per hour on a 24-foot-diameter circular path, pulling an average of 180 pounds. Multiplying it all out gives for the horse's work capacity $24\pi \times 180 \times 144/60 =$ 32,572 foot·pounds per minute, which James Watt rounded up to the familiar figure of 33,000, the equivalent of the 550 foot·pound per second that became the basis for the definition of the standard horsepower. This value of the horsepower is part of our customary system of measures.

Horses were also used in plowing soil. A typical horse could plow a furrow of a certain length in an hour. This length was the furrow-long, now known as the furlong, familiar to anyone who enjoys the "Sport of Kings."

Although James Watt (1736–1819) has been linked to the steam engine, Thomas Savery and Thomas Newcomen were the first to pump water from coal mines by means of heat engines. Their early machines generated a partial vacuum by condensing steam inside the machine's upright mounted cylinder. To this end, a jet of cold water was injected into the steam-filled cylinder so that atmospheric pressure on the piston's upper side forced it down into the vacuum below. The machines didn't have a flywheel but used a counterweighted rocker arm to act on the pump that the engine was driving. Their design was awkward. On one side there was complexity, on the other, inefficiency. Watt saw the obvious: namely, that using steam for its pressure showed more promise than using it to no better end than to make a vacuum. In all probability, though, problems in boiler fabrication were the stumbling block. Newcomen machines needed little pressure— just enough to blow steam into the empty cylinder—while James Watt's design required explosion-proof boilers. With the technology of those times, it was not an easy task to make them.

But that wasn't Watt's only problem. He had to design a mechanism for converting the forward and backward motion of the machine's piston into the rotation of a flywheel other than the crankshaft, which had already been patented. He also invented the flyball governor, which automatically regulated the speed of the engine. Most important is his concept of the double-action engine, in which steam is admitted by means of a slide valve alternately at one or the other end of the cylinder, doubling the output.

Such breakthroughs justify James Watt's title as the inventor of the steam engine, even if the basic idea of such machines had been around before. Watt's industrial-size engine, installed in 1777 at Smethwick, England, set a record of durability by operating for nothing less than 120 years until its removal in 1898. Thanks to their compactness, Watt's steam engines even provided marine propulsion, and in 1788 William Symington took a steam-powered catamaran across Dalswinton Loch in Scotland.

The concept of the horsepower, of 550 foot·pounds per second, suggests that a single horse could consistently pump one pound of water up from a 550-foot-deep shaft. We may equally imagine the lifting of 10 pounds from a 55-foot-deep well, or any other pair of numbers whose product is 550. Thus, we get for the amount P of horsepower needed for lifting W pounds per second of water from an H-foot-deep well:

$$P \,(\text{in HP}) = W \times \frac{H}{550}. \qquad (10.1)$$

Watt's horsepower actually exceeded the capacity of all but the strongest work horses. We will never know if that was accidental or the inventor-turned-entrepreneur's little trick for pleasing his customers; in any case, the measure made his steam engines perform consistently far above the buyers' expectations.

When the metric system took hold on the other side of the English Channel, people unwilling to let go of the eminently practical unit of horsepower coined their own metricated version. The horsepower became 75 meter·kilogram per second, the equivalent of 735.5 watts of power, while Watt's original horsepower equaled

745.7 watts. In comparison, a human, exercising at a steady pace, averages a measly 1/7 HP output.

Because of their common roots, both electrical and mechanical energy are expressed in identical units of measure. Mechanical work W, the product of force F and distance covered s is

$$W = F \times s. \qquad (10.2)$$

For work done in t seconds, the respective power P, or work per second, is given by

$$P = \frac{W}{t}. \qquad (10.3)$$

Combining the two formulas gives $P = F \times s/t$, and with $v = s/t$ for the velocity of displacement, we get for power:

$$P = F \times v. \qquad (10.4)$$

Think, for instance, of pushing a stalled automobile (Fig. 10.1) of 1500 kg mass up a 30° slope at the pace of one meter per second (3.6 km/h). Because sin 30° = 0.5, each meter of forward motion lifts the car upward by 0.50 meter, the work done being 0.5 × 1500 = 750 kfg·m. With the horsepower equal to 75 kfg/m·sec, that's 10 HP. A 10 HP engine could thus drive the car uphill at

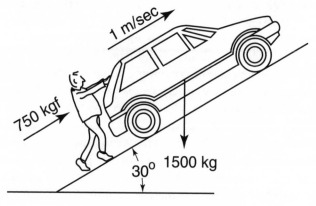

Figure 10.1 Laboring uphill

one meter per second, or if all engine powers failed, ten horses could (hopefully) do the job. Of course, a Porsche Targa with its typical 3.6-liter six-cylinder engine with an output of 320 HP could roar uphill at a speed exceeding 100 km/h. As 1 kgf = 9.80665 N, the HP is equivalent to

$$1 \text{ HP} = 75 \text{ m·kfg/sec} = 75 \times 9.807 = 735.5 \text{ N·m/sec}$$
$$= 735.5 \text{ J/sec},$$

and thus 1 HP = 735.5 W = 0.7355 kW. In all that, the watt can be understood as the amount of electricity needed to move a body (in the absence of friction) at the constant speed of one meter per second against a force of one newton (0.102 kgf).

Embedded in the 1:0.736 horsepower-to-kilowatt ratio lies a convenient rule of thumb for figuring the amount of electricity consumed by small electric motors. Their efficiencies usually resemble that same figure, so that a 1 HP electric motor can be expected to need approximately 1 kW of electricity, a 2 HP motor 2 kW, etc. However, larger machines, typically those from 10 HP up, have better efficiency ratings than those this rule is based on.

The Ampere: Basic Unit in Electricity

The French physicist and mathematician André-Marie Ampère was born in 1775 in Lyon, where his father was justice of the peace. In 1792, when the city fell to the forces of the French Revolution after a siege of two months, Ampère's father was arrested and sent to the guillotine. Ampère never fully recovered from shock. Nevertheless, he founded and named the science of electromagnetics, which is now commonly known as electromagnetism, which led to the law of electromagnetism that depicts the magnetic force between two electric currents.

This phenomenon of mutual attraction between a pair of electrical conductors (wires) has led to the definition of the ampere as the electric current that flows in each of two parallel conductors, spaced one meter apart, if they mutually attract each other with 2×10^{-7} newton of force per meter of conductor length. These forces become significant in power transmission lines with two, three, or four parallel conductors per phase, when short-circuit

currents generate enough mutual attraction for the strands to clash violently and damage each other, while bus bars in the inter-connecting substations get bent.

As happened with the units of meter and second, the ampere too has since morphed into a quantum-physical definition without any significant changes to its magnitude. Thus, the ampere has become the flow of 6.2424×10^{18} electronic unit charges (electric charge of the electron) per second, allowing for interesting con-clusions. With the electron's mass of 9.035×10^{-31} kg, and 31.536×10^6 sec to the year, the total mass of electrons channeled through a wire in one year's time by a *one ampere* electric current becomes

$$6.2414 \times 10^{18} \times 9.035 \times 10^{-31} \times 31.558 \times 10^6$$
$$= 0.178 \times 10^{-3} \text{ kg} \approx 0.2 \text{ gram}.$$

Thus, a typical 2200-watt room heater, consuming about 20 A of electric current, would pump throughout the year 20×0.178 gram $= 3.56$ grams $\approx 1/8$ ounce of electrons through its power cord.

Electric Charge

The quantum-physical definition of the ampere as the flow of 6.2414×10^{18} electrons per second also makes the electric charge of the unit of one coulomb (C) equal to the combined charge of 6.2414×10^{18} electrons. Long before humans learned how to generate electric current, charge was the only kind of electricity scientists could use. Charge shows up in our daily lives as static electricity, such as when shreds from plastic grocery bags stick to our fingertips, or when our hotel room's doorknob deals us a shock as we touch it after walking along a long, nylon-carpeted hallway. Lightning too is powered by electrically charged clouds.

There is mutual attraction between positive and negative charges and, conversely, repulsion of equal charges. The force F acting between the charges Q_1 and Q_2, placed at a distance R center to center (Fig. 10.2), is given by Coulomb's law as

$$F = k_0 \frac{Q_1 \times Q_2}{R^2}, \tag{10.5}$$

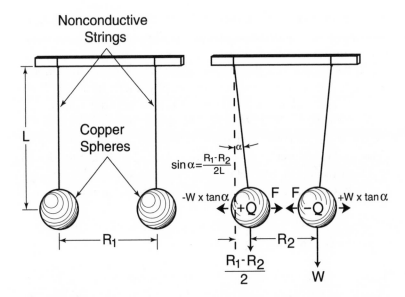

Figure 10.2 Measuring mutual attraction between opposite electric charges

which is analogous to Newton's universal law of gravitation. Setting the proportionality constant $k_0 = 1$ led to the earliest set of electrical standard units, comprising the electrostatic system with the unit of statcoulomb for charge. The original "Absolute System of Measures," also known as the centimeter·gram·second (cgs) system, the first that defined the gram as the unit of mass rather than of weight, included the electrostatic system of electrical units.

Had Charles Augustin Coulomb (1736–1806) worked in Great Britain or America, the unit of statcoulomb would have become the charge on each of two spheres, which mutually attract with the force of one pound. With the conversions

1 foot = 30.48 centimeters (cm), 1 pound
       = 453.6 grams (g), 1 pound force = 444,822 dyn,

Coulomb's law would read,

$$F = 444822 = \frac{Q_1 \times Q_2}{R^2} = \frac{Q^2}{R^2} = \frac{Q^2}{30.48^2},$$

which we resolve into

$$Q = 30.48 \times \sqrt{444822} = 20328.63 \text{ statcoulombs.} \quad (10.6)$$

Thus, the U.S. coulomb would have been 20,329 times weightier than the European coulomb, but that wasn't in the cards. A proprietary unit for electric charge was never established in America, and even the metric statcoulomb was short lived. Because electric currents can be measured with much higher precision than electric charges, the International System of Weights and Measures incorporated the ampere as its fundamental electrical unit.

Thus, the original approach to fixing these units was inverted: while the absolute system made the statcoulomb its fundamental unit and defined the statampere as the current of an electric charge of one statcoulomb per second, the International System established the value of the ampere and derived the unit of charge from it. This new coulomb, 2.99792×10^9 the magnitude of the statcoulomb, makes the proportionality constant k_0 in Eq. (10.5) quite different from unity. Its new value can be found by rearranging equation Eq. (10.5) into $k_0 = F \times R^2 / (Q_1 \times Q_2)$ and introducing the conversion factors 1 dyne $= 10^{-5}$ N, 1 cm $= 10^{-2}$ m, 1 gram $= 10^{-3}$ kg. This leads to

$$k_0 = \frac{10^{-5} \times (10^{-2})^2}{\left(\dfrac{1}{2.997924} \times 10^{-9}\right)^2} = \frac{8.98755 \times 10^{18}}{10^5 \times 10^4}$$

$$= 8.98755 \times 10^9 \text{ N·m}^2/\text{C}^2.$$

Therefore, Coulomb's law in the units of the International System reads as

$$F = 8.98755 \times 10^9 \times \frac{Q_1 \times Q_2}{R^2}. \quad (10.7)$$

In a setup like Figure 10.2, this equation gives the force (in newtons) between two spheres with identical charges as

$$F = 8.98755 \times 10^9 \times \left(\frac{Q}{R}\right)^2. \quad (10.8)$$

The capacity of a sphere for storing electric charge is proportional to its radius r, namely $\frac{1}{9} \times 10^{-11} \times r$. For instance, a sphere with $r = 10$ cm radius, charged to 100,000 volts, carries the electric charge of

$$Q = \frac{1}{9} \times 10^{-11} \times 0.10 \times 100,000 = \frac{1}{9} \times 10^{-7} \text{ C.}$$

If a pair of such spheres is spaced 25 cm (0.25 m) center to center, they attract each other (Eq. 10.7) with the force of

$$F = 8.98755 \times 10^9 \times \left(\frac{\frac{1}{9} \times 10^{-7}}{0.25} \right)^2 = 1.775 \times 10^{-5} \text{ N}$$

$$= \frac{1.775 \times 10^{-5}}{9.806} = 1.810 \times 10^{-6} \text{ kgf.}$$

That's about 2 mg of force and would be hard to measure in an elementary setup like Figure 10.2. But Charles A. de Coulomb developed, in 1784, a torsion balance sensitive enough for direct measurements of the minute mechanical effects of electric charge. A 1761 graduate from the Ecole du Génie at Mézières, Coulomb's work in engineering included structural design, soil mechanics, and the building of a new fort in Martinique in the West Indies. His work, titled "Théorie des machines simples," rendered him the Grand Prix from the Académie des sciences in Paris, which a few short years later was closed down because it was seen as a bourgeois institution by the revolutionaries. Retired to a country farmhouse, Coulomb developed his interest in electricity and magnetism; his experimental proof of the inverse square law became the basis of Poisson's mathematical theory of magnetism.

The reason we encounter such dishearteningly small forces of electrostatic attraction (Coulomb forces) lies in the low storage capacity of ordinary objects; save for bodies the size of a blimp, there simply is not enough surface area to harbor a sufficient number of electrons. By contrast, a 1000 μF capacitor, common in electronic circuits, charged to 200 volt, holds $1000 \times 10^{-6} \times 200$

$= 2 \times 10^{-1} = 0.2$ C. If such a charge could be stored on the spheres in Figure 10.2, their mutual attraction under otherwise unchanged conditions would become a sizable

$$F = 8.98755 \times (0.2/0.10)^2 = 35.95 \text{ N} = 3.67 \text{ kg} \approx 8 \text{ lb.}$$

In all that, let's not forget that electrostatic forces bind the electrons to the protons of the atomic nucleus and are thus essential for the existence of matter.

A Conversion Experience

We have seen that the use of the customary FPS (foot-pound-second) units leads to a "customary coulomb" of $Q = 30.48 \times \sqrt{444822} = 20328.63$ statcoulomb. Further, we know that the International System's coulomb equals 2.99792×10^9 statcoulombs. This gives for the relation

$$\frac{\text{SI-coulomb}}{\text{FPS-coulomb}} = \frac{2.99792 \times 10^9}{20328.63} = 147473.0.$$

For convenience, this conversion factor can be built into the proportionality constant of the equation for Coulomb's law if we multiply Q_{SI} by the factor above:

$$F = \frac{(Q_{SI} \times 147473)^2}{R^2} = 2.17483 \times 10^{10} \times \frac{(Q_{SI})^2}{R^2}.$$

In this case, Coulomb's law for use with electric charge in units of the International System, and all other variables in FPS units, reads

$$F = k_1 \frac{Q_1 \times Q_2}{R^2} = 2.1748 \times 10^{10} \times \frac{Q_1 \times Q_2}{R^2}. \quad (10.9)$$

As a quick check, let's figure in FPS units the force among a pair of (one coulomb) charges, spaced 3.28084 ft (one meter) from each other. Eq. (10.9) gives for this case

$$F = k_1 \frac{Q_1 \times Q_2}{R^2} = 2.17483 \times 10^{10} \times \frac{1 \times 1}{3.28084^2}$$

$$= 2.02048 \times 10^9 \text{ lbf,}$$

which with 0.224809 newton to the pound gives

$$F_N = \frac{2.02048 \times 10^9}{0.224809} = 8.98755 \times 10^9 \text{ N,}$$

identical with what we get from Eq. (10.7) for $Q_1 = Q_2 = 1$, and $R = 1$.

Altogether, these are examples typical for the development of units of measure. Either you start up with giving the unit in point a determined value and subsequently compute the proportionality constant in the equation for the pertinent law of physics, or you make that constant a priori equal to 1 and deduce the unit from the equation. Coulomb took the first approach in setting the value of the statcoulomb. The International System, having to deal with the predetermined unit of ampere-seconds for charge, ended up with setting the proportionality constant in Coulomb's law at a value quite different from 1.

Electric Energy and Potential

As water power is linked to the flow of water, so is electric power related to the flow of electric current (Fig. 10.3).

While Q kilogram (or liter) of water per second from a reservoir at H meter of altitude yields the power of

$$P = Q \times H \text{ m·kg/sec,} \tag{10.10}$$

the flow of J ampere of electric current, driven by the electrical potential E, yields

$$P = J \times E \text{ watt of power.} \tag{10.11}$$

Energywise, electric current (J) resembles the flow of water, and voltage (E) the "head" of the water supply, that is, the height of the reservoir or lake above the power-generating machinery.

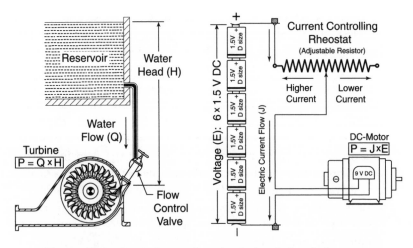

Figure 10.3 Equivalence of mechanical and electrical power

The unit of volt is named after the Italian count Alessandro Volta, professor of physics at the university in Pavia, Italy. The count was first to generate, in 1800, a steady electric current. Until then, experiments had mainly been done on static electricity.

Volta interspersed piles of plates, made from different metals, with pads soaked in acid or salt solutions. This marked the birth of dynamic electricity, as before then only occasional use had been made by current surges obtained from Leyden jars, the forerunners of today's capacitors.

The principle of such batteries had unwittingly been discovered by the Italian anatomist Luigi Galvani, who hung a dead frog from an iron hook and made its legs twist by touching their extremes with metal electrodes. This happened because the fluid retained in the frog's muscular tissues acted much like the electrolytic solutions in galvanic cells, where plates of unequal metals, submerged in electrolytic solutions, spontaneously develop a voltage difference relative to each other. To touch the frog leg with a piece of copper wire would have generated a voltage of 1.05 volt (V) relative to the iron hook the animal was hanging from. Nickel would have given a whopping 3.3 V. However, Galvani still clung to the concept that "animal electricity" explained such muscular

contractions, which was the popular theory of those times. Nevertheless, the word "galvanometer" is still used occasionally for instruments that measure the flow of electricity.

In the wake of Galvani's demonstrations, Napoleon Bonaparte's chief surgeon, Dominique Jean Larrey, demonstrated muscular contractions on freshly amputated human legs for an audience gathered principally to admire his speed. In the days before anesthesia, the swiftness, rather than the quality, was the measure of a doctor's surgical skills because the pain in a prolonged procedure would eventually kill the patient. If the chronicles are correct, Larrey could amputate a leg in thirteen to fifteen seconds, a figure that decades later the record holder of the time, Robert Liston, tried to top in vain: he couldn't do better than two minutes and a half. Liston, nominated by *Blackwood's Magazine* as "The Great Northern Anatomist," made his fortune as professor of surgery at the University College Hospital in London. On one occasion, the story goes, he tried so hard to beat his own record that he not only severed the patient's gangrened leg, but he accidentally amputated the fingers of his assistant as well as the coattails of a dignified spectator, who, terror stricken, dropped dead on the spot.

In a less bloodstained scenario, "volt" stems from the term "voltage" given for the electric potential of a source of electricity. Metric countries use the word "tension" instead.

In the analogy between liquid flow and the flow of electricity, voltage relates to water pressure (Fig. 10.3) and amperage to the flow of water per second. Such comparisons are helpful in understanding the basic laws of electricity but fail to account for induction, the generation of magnetic fields by electric currents. No power fields of any kind have ever been found surrounding the veins of flowing water, although some dowsers and health prophets still insist on their omnipresence in one form or another.

Resistance and Electric Current

Every electric conductor, from copper wire to the filament of an incandescent lightbulb, irrevocably converts electrical energy into thermal energy (heat) and thus constitutes an electrical *resistor.*

The energy converted (often called "lost," though energy cannot simply vanish) causes an equivalent voltage drop between the ends of the resistor.

If the resistance of a computer's 6-foot power cord causes the voltage between the wall socket and computer to drop by ΔE = 0.1 V, we intuitively expect a 12-foot cord, connected to an identical load, to cause twice that voltage drop, namely 0.2 V. Likewise, 0.3 V would be lost in an 18-foot power cord, and so on. The voltage drop ΔE over a resistance R is thus proportional to the resistance: $\Delta E \propto R$. It shouldn't be a surprise that the voltage loss over the resistance of the cord is likewise proportional to the electric current J: $\Delta E \propto J$. Both expressions can be combined into the statement

$$\Delta E \propto J \times R \text{ or } \Delta E = K \times J \times R,$$

where K is the proportionality constant. With $K = 1$, the above equation becomes Ohm's law, stating the relation between the resistance R of an electric conductor, the applied voltage E, and the electric current J:

$$E = J \times R \qquad\qquad (10.12)$$

$$J = \frac{E}{R} \qquad\qquad (10.13)$$

$$R = \frac{E}{J}. \qquad\qquad (10.14)$$

Eq. (10.14) contains the definition of the ohm (Ω) the unit of electrical resistance, as the resistance of a conductor that, connected to a one-volt source, allows a one-ampere current to flow. Not everything, however, is an ohmic conductor. The simple relation $E = J \times R$ does not hold for diodes, valves, and transistors, which are not governed by this law.

Resistance and Electric Power

Now, because $P = E \times J$ and $E = J \times R$, the amount of energy P dissipated per second in a resistive circuit is

$$P = J \times R \times J = J^2 \times R. \qquad (10.15)$$

Introducing $J = E/R$ from Eq. (10.16) yields further

$$P = \frac{E^2}{R} \text{ and} \qquad (10.16)$$

$$R = \frac{E^2}{P} \qquad (10.17)$$

For instance, the resistor in a 2400-watt electric water heater element, operated at 220 volts, needs to have the resistance of $R = E^2/P = 220^2/2400 = 20.17 \ \Omega$. We need not worry about the 0.17 ohm; generous tolerances in resistors are not uncommon. Resistors found in radios and televisions are usually of the $\pm$ 10% variety. Commercial precision resistors are kept within $\pm$ 1% of their rated resistance.

Electrical Resistance in Transmission Lines

In high-voltage power distribution, two or more relatively thin cables in parallel are often preferable to a heavier single conductor. That's because wind-induced (Aeolian) vibrations are more likely to cause cracks in thick cables than in the intrinsically more flexible thinner ones. Moreover, the thinner the conductor, the higher its ratio of surface area to cross-sectional area, and thereby its heat dissipation, which counteracts the trend for high-current loads to heat up the cable. While the energy dissipated as heat in an electric water heater, range, dryer, or soldering iron is useful, heat generated in electrical conductors is wasted. Reducing such waste in transmission lines that span over hundreds of miles is of paramount importance.

The specific resistance ρ for a given material is the resistance of a wire of one square millimeter cross section and one meter of length of that material. Conductance is the reciprocal of specific resistance and should thus be expressed in Ω^{-1}. However, the term mhos (from "ohm," read backwards and a plural "s" added) and the International System's unit of conductance, the siemens (symbol S), are used.

With ρ = 0.016 Ω/m·mm^2, pure silver is the best conductor, followed by electrolytic copper (0.0172), and gold (0.0244). Next in line is aluminum, with ρ = 0.0279 Ω/m·mm^2. Because of their cost, precious metals are out of the question as conductive materials, but copper and aluminum can serve the purpose nicely. Copper is the primary choice for internal installations, domestic as well as industrial, thanks to its conductivity, its mechanical and fatigue resistance, and its ease of soldering. Compared to aluminum, copper does not settle under a static load, so that connectors with copper wires, once tightened, remain in place and don't need to be retightened. However, copper's high density of 8.9 kg/dm^3, combined with its high price, makes aluminum the material of choice for long-distance power transmission lines.

That the electric conductivity of aluminum amounts to only 60% that of copper is more than compensated for by aluminum's low density of 2.75 kg/dm^3. Relative to cable weight, the conductivity of aluminum turns out to be even better than that of copper, because copper is 8.9/2.75 = 3.24 times heavier than aluminum, but only 0.0279/0.0172 ≈ 1.6 times more conductive. This means, pound for pound, that aluminum cables can carry 3.24/1.6 ≈ 2.0, or about twice the electric current of copper cables.

The sole shortfall of light-metal cables is their low fatigue resistance, which limits line tensioning to 20% of their ultimate tensile strength. This weakness led to the development of ACSR (Aluminum Conductor Steel Reinforced) cables, which consist of a steel cable for mechanical strength, covered by several layers of helically wound, relatively thin aluminum wires.

Sources of Electric Power

Flashlight batteries produce 1.54 volts of energy. Automobile lead-acid batteries deliver the familiar 12.6 volts DC of automotive systems by connecting eight cells of 1.57 V each in series. Utilities supply industrial power as three-phase 480 volt alternating current (AC), as well as the familiar single-phase domestic 120 VAC (volts AC).

In direct current (DC) systems, one wire is always negative and the other is always positive. In AC installations, polarities invert

sixty times per second because the alternating current is generated by rotating a coil of wire (armature) through a magnetic field, so that the voltage, induced in those windings, changes direction at every half turn. What's more, this happens in ways that make the magnitude of the output voltage follow a sine function.

Distribution

Direct current, which dates back to the days of Volta's galvanic cells, was Thomas Edison's source of electricity in developing the lightbulb. Thus, the system of electric lighting he introduced in New York City in 1882 operated on DC. Over the following years, Edison became the strongest proponent of DC-powered utility networks.

But dynamos producing DC electricity have inherent problems. Electric current, generated by induction in the machine's rotating armature, is discharged via a commutator and brushes, wherein the risk of sparking limits operating voltages to about 1500 volts. This is far too little for power distribution over distances greater than a few city blocks. As the internal resistance of supply lines brings the voltage down, power $P = E^2/R$ suffers still greater losses. If the voltage drops by ΔE from E to $E - \Delta E$, power shrinks proportionally to $(E - \Delta E)^2$, that is, $E^2 - 2\Delta E + \Delta E^2$. The last term is negligible in comparison to E, so that power losses amount to nearly double the voltage drop. If voltage goes down by, say, 10%, power will do so by 20%. That's why lights dim when a heavyweight consumer connects into the network.

Nevertheless, Edison held on to his belief in DC, even when Nikola Tesla, then an employee in Edison's enterprise, developed the alternating current generator, which uses simple slip rings instead of a commutator. Frustrated by Edison's stubbornness in sticking to DC, Tesla accepted a job offer from George Westinghouse, a one-time Pittsburgh railroad entrepreneur. His company, Westinghouse Electric, subsequently became the standard bearer of alternating current technology.

In the bitter battle with Westinghouse Electric that ensued, Edison's principal argument rested on safety issues. For equal power output, he correctly argued, AC peak-to-peak voltage must

be made 2.83 times higher than DC voltage, something Edison claimed was a threat to public safety. He made his point dramatically by publicly using AC shocks for the electrocution of stray dogs and cats that had already survived exposure to apparently identical DC voltages. Even Topsy, a Coney Island elephant, accused of having trampled three people to death and therefore a candidate for execution, died under Edison's AC treatment. In the heat of battle, the introduction of the electric chair in the U.S. penal system became an opportunity to forge such terms as "condemned to the Westinghouse," as if old George were of the same fiber as Dr. Guillotine, whose device removed 16,000 aristocratic heads during the French Revolution.

In the end, Edison didn't stand a chance. In 1893, Westinghouse "outlighted" him at the World's Columbian Exposition in Chicago with a flood of AC-powered lights of a richness never seen before. In industry, too, the alternating current prevailed because AC motors and generators are less complex and more durable than their DC counterparts, and alternating voltage can be converted up or down in transformers of simple design (Fig. 10.4).

The ease of converting voltage is important because long-distance transmission of electric power is possible only at very high voltages. This is so because for a given measure of power, Eq. (10.11), $P = J \times E$, makes current and voltage inversely proportional. The higher the voltage, the lower the flux of current, and with it, the gauge of conductors for the job. For this reason, voltages as high as 660,000 VAC have become typical for country-wide networks.

With the peak-to-peak AC voltage being $2 \times \sqrt{2} = 2.83$ times higher than the effective voltage, 660 kV AC amounts to 1868 kV peak to peak, or close to 2 million volts, which marks the limits of the insulating capacity of the air between cables. Still higher voltages tend to cause glimmering corona discharges, and even sparking.

This allowed for a surprising comeback of direct current in power transmission, because effective and peak-to-peak voltages are identical in DC systems. Direct current transmission is thus being employed where the extent of the transmission lines justifies the additional costs of DC voltage conversion gear, which is considerably more complex and costly than its AC counterpart.

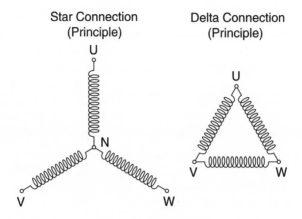

Star Connection
(Principle)

Delta Connection
(Principle)

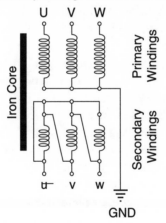

Three-Phase Power Transformer
Primary Windings in Star Connection
Secondary Windings in Delta Connection

Figure 10.4 Transformer configurations and motor windings in three-phase AC applications

Electric Power Abroad

Unlike the American 480/240 volt standard, European utilities work with three-phase, 380 volt alternating current, using the corresponding 220 VAC phase-to-ground voltage for domestic supply (Fig. 10.4). Compared to the U.S.'s 120 VAC standard,

the higher voltage results in lower electric currents and thus allows for lighter wiring, but it heightens electric-shock hazards.

The √3 : 1 Bonus Ratio

Because of the three-phase alternating current's intrinsic ratio of √3 :1 between phase-to-phase and phase-to-ground voltages, distribution transformers with 380 volts phase to phase in star configuration (Fig. 10.4, left) can supply the domestic voltage of 220 volts between neutral and any of the three phases without separate windings or taps. Transformers with primary windings in star and secondary windings in delta (Fig. 10.4, right) command an inherent voltage reduction of 1.732:1 on top of the transformer's basic voltage conversion ratio.

On the other end, three phase AC motors often allow for user configuration of their windings, either in star for industrial use on 380 VAC (Fig. 10.4, left) or in delta (Fig. 10.4, center) for running them from 220 V supply lines.

The AC Frequency Barrier

In alternating current technology, frequency, the number of polarity swaps per second, also known as cycles per second, is expressed in hertz (Hz) in all systems of measures, but European utility grids carry 50 Hz AC versus the 60 Hz in the United States. This, along with the different voltage levels between here and there, makes a worldwide unified utility network virtually impossible. Besides, installing ocean-crossing electric power lines between continents would not be economically feasible.

Continental European electrical outlets use round pins only, so that unknowing American tourists can't insert their 120 V shavers or hair dryers into 220 V power outlets. U.S.-made electric clocks that synchronize with the utility frequency still lag behind by 17% when plugged (via step-down transformer and plug adapter) into European outlets. Likewise, European clocks used in the Americas run ahead of their native siblings by 20%.

Surprisingly, standard three-phase industrial motors (squirrel-cage induction motors) have often been used on both networks so long as their rated voltage could be matched approximately

with local standards. Like their timekeeping counterparts, 50 Hz motors run faster on 60 Hz power grids. The most common type, the four-pole induction motor, rated typically at 1450 rpm in 50 Hz installations, goes up to about 1740 rpm if fed by 60 Hz power. However, since impedance (a motor's apparent resistance), too, is frequency dependent, the "transplanted" motors tend to overheat. Yet, such replacements are not entirely uncommon and mildly successful in countries still caught in the earlier stages of industrialization.

The Efficiency Deficit

Conversion of thermal to mechanical energy happens at very low efficiency, typically about 30% in modern power plants. Adding some 5% for losses in transmission, voltage transformation, and distribution, brings the overall efficiency at domestic outlets down to 25%. Therefore, if reconverted into heat, electricity releases no more than one quarter, and sometimes even less, of the thermal energy it took to generate it.

Thermoelectric power plants generate from one pound of Diesel oil 1.39 kWh of electricity, the equivalent of 4750 BTU. But combustion of the same one pound of Diesel oil in a room heater provides 19,000 BTU. Therefore, electric room heating, the cleanest and most convenient manner for keeping warm, is also the most expensive one.

A more rational approach is to elevate the temperature of an existing heat reservoir, typically ambient air, to a higher temperature level by reverse cycling an air conditioner, which in that case becomes a heat pump. Surprisingly, the efficiency of heat pumps may exceed 100%.

· 11 ·

Compound Units

Ancient Egyptians used the length of their forearm from the (bent) elbow to the tip of the middle finger in ways we use a yard-stick to measure distances. Sometime around 3000 B.C., though, authorities had the "Royal Cubit Master" carved out of a bar of granite and made it the nationwide standard of length. Likewise, the weight and mass unit of "kite" appeared in quite arbitrary magnitudes, and the unit of time got linked to the length of day-light—a very unfortunate choice indeed, because it varied throughout the year.

Had such random choices continued into present days, we would be left with an endless set of units with no links to one another. Though inconvenient, such a collection could still sur-vive in present-day science and technology, but we fare better if we minimize arbitrary choices and derive new units from the existing ones. The International System of Weights and Measures (SI) has been configured by deducing "compound units" from a small set of arbitrary, "basic" units through the appropriate laws of physics. Highest visibility in that category went to the unit of "newton" (N), the SI unit of force derived from Newton's second law: $F = m \times a$. Herein, a is the acceleration in meters per second squared of a mass m, acted upon by the force F. Unit force would thus accelerate the mass of 1 kg by 1 m/sec^2. In free fall, an object driven by the force of its own weight acceler-ates by g = 9.80665 m/sec^2, so that the force for a = 1 m/sec^2, the equivalent of the newton, becomes $1/g$ = 1/9.80665 = 0.10197 kgf, where kgf is the symbol for the weight of one kilo-gram of mass.

In all that, we took the units of meter, kilogram, and second for granted. Much like the axioms of mathematics, there is nothing from which those basic units could be deduced. They were chosen for convenience: the second from the combined length of day and night, and the original meter as the ten-millionth of the quarter meridian. Conversely, the prefix "kilo" in the symbol for the unit of mass hints at its derivation from the one thousand times lighter "gram." On the other hand, the definition of the gram as the mass of one cubic centimeter of water brandishes the gram as deduced from the specific gravity of water, arbitrarily set equal to 1. Opponents of the metric system used this arbitrariness to question the kilogram's qualifications as a basic units, but that's a merely philosophical issue. Once we rank the kilogram among basic units, its process of generation becomes irrelevant.

The three basic units of mechanics aside, another four are needed to cover the fields of thermodynamics, optics, electricity, and physical chemistry. We recall them from preceding chapters as the degree kelvin (K) for thermodynamic temperature, the candela (cd) for luminous intensity, the ampere (A) for electric current, and the mole (mol) for the amount of substance. Finally we have the "Supplementary Units," comprising the units for plane angle and solid angle, called the radian and the steradian.

Other compound units include the coulomb (electric charge), the volt (electric potential), and the ohm (electric resistance), obtained from the equations that link the watt·second (joule) and the ampere. They share the privilege of proper names with only a selected few in the community of compound units, as shown in Table 11.1. All of these symbols can be used with the familiar prefixes for indicating multipliers. We understand the broadcast band of radio frequencies as covering 530 kHz to 1650 kHz, rather than 530,000 to 1,650,000 Hz, though both expressions are correct. Likewise, we scan the FM band between 88 and 108 MHz, or 88 × 10^6 to 108 × 10^6 Hz. Governments sell bandwidth in the gigahertz (10^9 Hz) range for gigadollars to the information industry, which envisions hard drives and memory chips with terabites (10^{12} bites) of storage capacity. Mechanical engineers routinely use the millimeter, of 10^{-3} or 0.001 meter, and, where precision really matters, as in the production of ball and roller bearings, the

Table 11.1 Compound Units

Field of	Physical Quantity	Name of Unit	Symbol	Dimension
Mechanics	force	newton	N	$m{\cdot}kg/s^2$
	work	joule	J	$m^2{\cdot}kg/s^2$
	power	watt	W	$m^2{\cdot}kg/s^3$
	pressure & stress	pascal	Pa	$kg/m{\cdot}s^2$
Electricity	electric potential	volt	V	$m^2{\cdot}kg/s^3{\cdot}A$
	electric resistance	ohm	Ω	$m^2{\cdot}kg/s^3{\cdot}A^2$
	electric conductance	siemens	S	$s^3{\cdot}A^2/\,m^2{\cdot}kg$
	electric charge	coulomb	C	$s{\cdot}A$
	electric capacitance	farad	F	$s^4{\cdot}A^2/\,m^2{\cdot}kg$
	inductance	henry	H	$m^2{\cdot}kg/s^2{\cdot}A^2$
	frequency	hertz	Hz	$1/s$
Magnetism	magnetic induction	tesla	T	$kg/s^2{\cdot}A$
	magnetic flux	weber	Wb	$m^2{\cdot}kg/s^2{\cdot}A$
Optics	luminous flux	lumen	lm	$cd{\cdot}sr$
	illuminance	lux	lx	$cd{\cdot}sr/\,m^2$

micrometer ($1\ \mu m = 0.001$ mm $= 10^{-6}$ m). Their colleagues in electronics solder capacitors of picofarads (1 pF $= 10^{-12}$ F) into their tuners and use microfarads ($1\ \mu F = 10^{-6}$ F) in filter circuits to smoothen DC power.

The following table lists the most common of those prefix multipliers:

Prefix	Symbol	Multiplier
tera-	T	10^{12}
giga-	G	10^9
mega-	M	10^6
kilo-	k	1000
hecto-	h	100
deca-	da	10
deci-	d	0.1
centi-	c	0.01
milli-	m	0.001
micro-	μ	10^{-6}
nano-	n	10^{-9}
pico-	p	10^{-12}

Note the importance of the case of the letters in symbols. For instance, the prefix m means milli-, but M stands for mega-. However, the name of a unit always starts with a lower case letter. Symbol abbreviations have no plural: 100 mm, not 100 mms. However, the spelled-out words can appear in plural when grammatically required: "They live twenty kilometers from our home"; however, one would say "I use an 8-millimeter drillbit with this tap."

Breaking the Rules—Again

Official FPS and SI units for velocity and acceleration should be part of our daily life, but few people would want their speedometers calibrated in feet/second, or on imports, meters/second. And highway signs spelling out any units other than miles per hour, or in metric countries kilometers per hour, would—to say the least—lack popularity.

There is nothing wrong with using such arbitrary units with linear equations, as in $s = v \times t$ for the distance covered in time t by a vehicle traveling at velocity v. Insert a velocity in meters per second to give you the distance traveled in meters; the velocity inserted in miles per hour will automatically yield a distance in miles.

But problems can arise if we need to convert between units, especially when it comes to second- or higher-order equations such as $s = a \times t^2/2$ for accelerated motion. For instance, if we mistrust the odometer of a car said to speed up to $v_e = 108$ km/h in $t = 10$ sec, we can check it by measuring the distance s, which the vehicle covers during the first 10 seconds of acceleration at full throttle.

But if we carelessly figured $a = 108/10 = 10.8$, we would get the obviously impossible figure of $s = 10.8 \times 10^2/2 = 540$ km. Our mistake was to be inconsistent. We should be converting 108 km/h to meter per second, and get $108 \times 1000/(10 \times 3600) = 3$ m/sec^2, and thereby $s = 3 \times 10^2/2 = 150$ m. Sometimes such "minor mistakes" can be fatal. It was a failure to convert from metric to imperial units that caused NASA to lose a Mars lander.

Faithful to Standards—After All

Sometimes it's easier to work with relative units. If you have a geostationary communication satellite and need to work out the force

on it, one relies on Newton's law of universal gravitation: $F = G \times m_1 \times m_2/R^2$. This requires looking up the value of the gravitational constant, the mass and the radius R_E of our planet, the weight of the satellite, and the radius of its orbit. The relative force on the satellite, compared to its weight on Earth, is

$$G\,\frac{m_1 m_2}{R^2}\Big/G\,\frac{m_1 m_2}{R_E^2} = \left(\frac{R_E}{R}\right)^2.$$

So, if the satellite is 42,162 km from the Earth's center and the Earth's radius is 6377 km, the force is $6377/42162 = 1/43.7$ of that on Earth, or 1/43.7 of the satellite's weight. This ratio will hold whether you use N, kgf, or lbf to express the force, or feet, cubits, furlongs, or kilometers for the radii of Earth and of the satellite's orbit.

Useful Tools

Errare humanum est—to err is human—is wisdom from ancient Rome. It has made its way unscathed right down into the information age, where it resurfaces under the term "human error" when apologists talk away otherwise unexplainable blunders. But a systematically developed system of measures presents ways to catch such human errors before they get a chance to develop their full potential of damage.

Though a dimensional check of equations does not prove them right, it definitely shows when they are wrong. For instance, if we mistakenly figured the formula for accelerated motion as $s = a/2 \times t^3$, the dimensional test would result in the contradiction $m = m/s^2 \times s^3 = m \times s$. In the course of extensive deductions, such tests save time by keeping us from endlessly following a course flawed by past mistakes.

Likewise, the errors in the former example of an accelerating automobile would have shown up in a dimensional analysis, as follows. Introducing the mistaken units in $s = a/2 \times t^2$ leads to $s = $ [km/(h·sec)] $\times$ sec^2 = km $\times$ sec/h, an impossible combination of two different units of time. We should have stopped right then and there, but once we make a mistake, that formula leads the way toward correcting it. With 1000 meters to the kilometer, and

3600 seconds to the hour, the right side of the equation becomes
1000/3600 = 1/3.6. Multiplying this with the "wrong" result of
540 gets us the correct value of 540/3.6 = 150 m.

Speaking of frequent mistakes, let's keep in mind that the term
"second squared" in the dimension of acceleration makes the conversion factor from "acceleration per second" into "acceleration
per hour" not 3600, but rather $3600^2 = 12.96 \times 10^6$.

Conceptual Formulas

Dimensions are handy tools to guess equations that describe physical processes. Physicists often term them "back of the envelope"
calculations—rough estimates that give some clue as to what the
"real" equation might be. For instance, we may try to deduce the
equation for the velocity v of water from an elevated reservoir as
follows.

Initially, we reason that the water's exit velocity must be
related to h, the height of the liquid level in the reservoir and to
the gravitational acceleration g, normally 9.80665 m/sec^2.
Hence, we start with the statement

$$v = k \times f(g \times h),$$

where f stands for "function of . . . ," and k for a proportionality
constant. Substituting the terms of this formula with the appropriate units makes it

$$\frac{m}{s} = f\left(\frac{m}{s^2} \times m\right) = f\left(\frac{m^2}{s^2}\right).$$

Only if f stands for square root will the two sides of this equation
become identical:

$$\frac{m}{s} = \sqrt{\frac{m^2}{s^2}} = \frac{m}{s}.$$

Our concept formula for the water's exit velocity thus becomes

$$v = k \times \sqrt{g \times h}.$$

The law of preservation of energy makes k equal to the square root of 2, so that $v = \sqrt{2gh}$, a result known as *Torricelli's theorem,* incidentally the same as the equation for the terminal velocity of a free-falling object, $v_e = \sqrt{2gh}$. Thus, the speed of liquid flow equals that of a liquid element descending in free fall from the height of the liquid surface in the reservoir.

Applications of the free-fall formula are farther reaching than the distance to the closest water tower. For instance, the wagons descending the track of the 225 ft (68.6 m) high "Steel Phantom" roller coaster in Kennywood Park, Pennsylvania, should speed up to $v_e = \sqrt{2 \times 9.806 \times 68.6} = 36.7$ m/sec, or $36.7 \times 60 \times 60/1000 = 132.1$ km/h or still $132.1/1.609 = 82$ mph.

Friction makes real-world velocities lag behind that value, but still, roller coaster designers beware: according to the formula for centrifugal acceleration, $a_c = v^2/R$, passage through a loop of, say, a 20 meter diameter ($R = 10$ m) would subject riders to centrifugal forces of $a_c = 36.7^2/10 = 134.6$ m/sec^2, or expressed in relation to gravitational acceleration, g: $134.6/9.807 \approx 14$ g, exceeding even the 8–10 g in a fighter pilot's training program.

Most roller coasters are laid out to limit the inertial forces on passengers to a mere 3.5 g, but that still makes them feel 3.5 times heavier than their real weight. What a blow for weight watchers! The minimum radius for the loops of the "Steel Phantom" can be found from the carts' velocity of 36.7 m/sec by rearranging $a_c = v^2/R$ into $R = v^2/a_c$, which gives for $a_c = 3.5$ g the smallest permissible radius of a loop as

$$R_{min} = 36.7^2/(3.5 \times 9.806) = 39.25 \text{ m}.$$

Roller coasters have come a long way since their forerunners, the fifteenth-century ice slides in St. Petersburg, Russia, earned them the nickname "Russian Mountains," except in their homeland, where they became "American Mountains." Whatever their names, demand for the thrills they provided knew no frontiers, neither geographic nor psychological. In 1804, not even an accident on the Paris roller coaster called "The Incredible Scream Machine" dampened the public's enthusiasm, and instead of scaring people away, enhanced the ride's popularity.

As we stand enticed by the standard meter's relation to the meridian quadrant, later the wavelength of the orange-red light from the krypton-86 isotope, and ultimately the speed of light, let's not neglect the good old British foot's nature-based virtues. Line up as many feet as there are days in the year, 365.242, and you have the length of one degree of equatorial longitude in units of 1000 feet. Check this out by converting this figure, with 0.3048 meter to the foot, into 0.3048 × 365.242 = 111.325 km, and compute the Earth's circumference by multiplying by 360 degrees, and you get 111.325 × 360 = 40,077 km, a mere 2 kilometers (less than one and a quarter miles) off the latest modern measurements of 40,075 kilometers. And they say our forefathers didn't know the size of their planet!

· 12 ·

Invasion of Aliens
and Nihilists

Definitions of unit quantities, such as the foot, pound, meter, kilogram, and second, usually monopolize the opening statements of physical sciences teaching. Long before audiences become familiar with the laws of physics in general, their minds digest a predetermined system of weights and measures, whether it's ours or the International System of Weights and Measures (SI).

Familiarity, of course, breeds contempt. In our case, working with a well-established, traditional set of units often makes people dismissive of others. So-called metric countries despise our use of inches, feet, and pounds, while Americans feel frustrated by their country's successive, albeit hardly perceptible, edging toward metrics. But all the arguments pro and contra in this two-centuries-long battle could be condensed into a few simple statements. The metricated ask why the United States, Burma, Brunei, and North and South Yemen don't follow the rest of the world and adopt the metric system. The world would benefit from such a unified system of measures.

Americans might reply "if it ain't broke, don't fix it."

Why should American industry incur significant costs in retooling if they don't stand to gain significant benefit?

A popular song from nineteenth-century Europe, when the battle for the metric system was raging, sums up the feelings prevalent back then and—I suspect—occasionally today as well. It's called "A Pint's a Pound the World Around" and goes as follows:

> Then down with every "metric" scheme
> Taught by the foreign school.
> We'll worship still our Father's God!
> And keep our Father's "rule"!
> A perfect inch, a perfect pint,
> The Anglo's honest Pound,
> Shall hold their place upon the earth,
> Till Time's last trump shall sound!

In the heat of battle, the combatants seem unaware that the fundamental laws of nature had been expressed long before reliable standards came into existence.

Twenty-five centuries ago, Archimedes, the ancient Greek philosopher, felt his body's buoyancy while soaking in a bathtub, which ultimately led to his well-known statement: "The loss of weight of a body immersed in a liquid equals the weight of the liquid displaced by the body." Legend has it that the usually dignified Athenian's excitement of discovery made him forget his state of undress as he burst onto the streets stark naked, clamoring: "Eureka—I found it!"

Characteristically for his times, Archimedes' law of buoyancy is devoid of any units of measure, and it survived into our times not *in spite* but rather *because* of that.

Archimedes' research was commissioned by Hieron II, the Tyrant of Syracuse, who suspected the gold content of a newly crafted crown was less than what the royal goldsmith had received to make it. Gold was then routinely alloyed with silver to improve strength and hardness, and an experienced goldsmith must have been able to guess the composition of such an alloy by its color: from the typical yellow of pure gold, to the silvery white of a 44/56% gold/silver alloy (Fig. 12.1).

Archimedes, I'm glad to report, didn't satisfy himself with guesswork. For starters, he built what has since been christened the "Archimedean Scale" (Fig. 12.2) for finding a body's buoyancy. Using water in a cylindrical container in lieu of weights, he could figure the object's density relative to the density of water from $H_1/(H_1 - H_2)$. That got him the density of gold, silver, and—most important—the crown. But he still had to do more math to get the formula

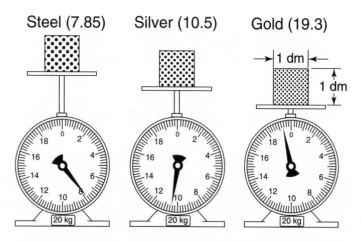

Figure 12.1 Specific gravity of steel, silver, and gold

$$W_{\mathrm{G}} = \frac{D_{\mathrm{c}} - D_{\mathrm{s}}}{D_{\mathrm{G}} - D_{\mathrm{s}}}$$

for the weight W_{G} of the crown's gold content from the densities D_{C} of the crown, and D_{S} and D_{G} for silver and gold, respectively. For instance, if the density of the crown had been weighed out as $D_{\mathrm{C}} = 16.0$, its gold content was

$$W_{\mathrm{G}} = \frac{16.0 - 10.5}{19.3 - 10.5} = 0.625, \text{ or } 62.5\%.$$

In our days, gold content is measured in the unit of carat, which like a percentage is dimensionless. Pure gold (100%) has 24 carats, a 75% alloy is marked $0.75 \times 24 = 18$ carats, etc., and the crown in our example would have been of $0.625 \times 24 = 15$ carats. By contrast, the emotionally charged carat number in engagement rings stands for the diamond's weight, equaling 0.2 gram per carat.

The chronicles lack detailed information on Archimedes' use of his knowledge, but as the intrepid goldsmith ended his life on the royal scaffold, we can believe that Archimedes uncovered some "missing mass" other than the one missing in today's concept of a "closed universe."

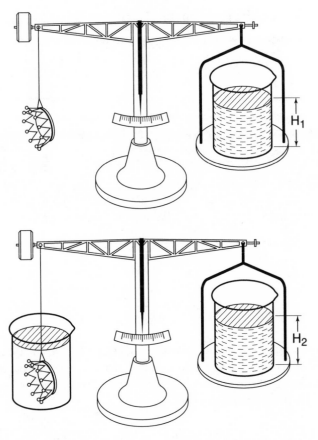

Figure 12.2 Archimedes' scale for weighing specific gravity. Submerged in water, the crown weighs less by the amount of liquid it replaces.

Archimedes' many scripts include the analysis of far more complicated problems, such as the areas of conic sections; the volumes of solids delimited by conic sections (Fig. 12.3); exponential notation of numbers, and even a rudimentary version of integration. In all that, Archimedes didn't think in pounds, kilogram, feet, meters, or, for that matter, any kind of units. Just imagine his thesis of buoyancy expressed in units current then in ancient Greece: "An object fully submerged in water loses $7\frac{10}{19}$ minas of weight for every khoes (1/12 amphora) of its volume." Would we remember Archimedes' principle in this format and make use of it in our

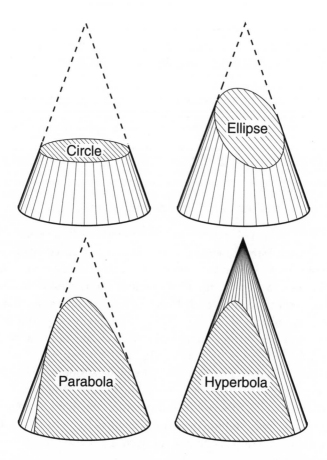

Figure 12.3   Conic sections

day? If so, the theorem's accuracy would be linked to the no-doubt-questionable degree of precision of the volume and weight standards of antiquity.

While Archimedes' version is absolute, like a mathematical statement, the version linked to units can never do better than the units with which it is expressed.

Even with the metric system, some uncertainty would persist. The statement "A body submerged in water loses one kilogram of weight for every cubic decimeter of its volume" sounds true enough. Yet, the latest update on the kilogram standard found its weight equivalent to 1.000028 liters of water (at 4 °C), rather

than exactly one liter, which was the desire and intention of the founding fathers of metric standards. This means the phrase "every cubic decimeter" in the above statement should be replaced by "every 1.000028 cubic decimeters," and that again could change with further improvements in the precision of our measuring instruments and methods.

Only because the original version of Archimedes' theorem contains no units at all, it is absolutely accurate and has survived unscathed into modern times.

A Heavenly Freedom from Units

Two millennia had passed since Archimedes' discoveries, when the German astronomer Johannes Kepler presented his famous laws of planetary motion—again without any reference to units:

Kepler's first law (Fig. 12.4) defines the orbits of planets as ellipses with the Sun at their focus. It is inherently unit independent, as is his second law (Fig. 12.5): "The areas swept out by the Planet-Sun line during equal time intervals are of equal magnitude." Likewise, his third law (Fig. 12.6) is expressed without reverting to units: "The squares of the planets' sidereal periods are proportional to the cubes of their orbits' major semi-axes."

While systems of weights and measures changed often and in many ways since Archimedes' and Johannes Kepler's times, their laws, conceived dimensionless, survived in the ways in which they were conceived.

Sir Isaac Newton, who developed the law of gravitation from Kepler's laws, made the question of units the least of his worries. As an undergraduate, he reports that an iron ball falls freely through "one hundred braces florentine in five seconds of an hour." Clearly, his understanding of gravity was not handicapped by the lack of an international system of units, or in this case, imperial units. Purists can calculate how far a ball falls in 5 seconds to show that "a brace florentine" is about 1.23 meters.

Astronomers' Bible

Historically, Kepler's discovery marks the break with fifteen centuries of Aristotelian philosophy. Aristotle regarded the circle as

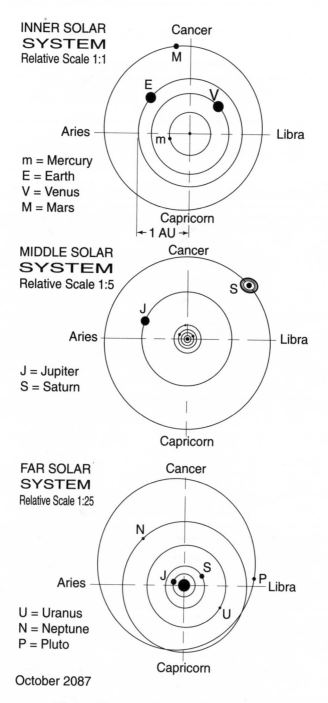

Figure 12.4 Elliptical planetary orbits

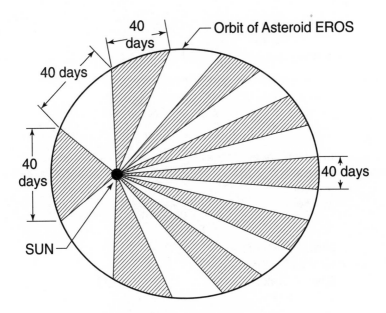

All segments are of equal area

Figure 12.5 Kepler's second law

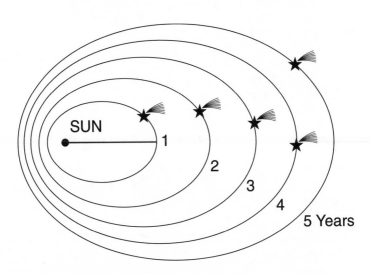

Orbits of fictitious comets of 1 to 5 years' revolution

Figure 12.6 Demonstration of Kepler's third law

the geometrical figure of utmost perfection, and, as the heavens were perfect, the circle must be the generating element of all celestial motions. Rather than belittle such ancient wisdom, we moderns should recall that from Later Antiquity and all through the Middle Ages, the concept of circular motions led to accurate predictions of the positions of the Sun, the Moon, and the planets, and that accurate calendars were compiled and solar and lunar eclipses were predicted correctly. This is no mean feat, especially as they regarded the Sun as traveling around the Earth.

Our knowledge hereof stems from the *Almagest*, an encyclopedic volume from second century A.D. astronomer and mathematician, Claudius Ptolemy. It describes the apparent orbits of the Sun, Moon, and planets as epicycles, curves generated by the superposition of several uniform circular motions—circles upon circles. Ptolemy's works include Hipparchus's catalog of over one thousand stars, whose precision enabled astronomers of the eighteenth century to detect the proper motion of some "fixed" stars for the first time. This was one of the earliest indications of the vastness of the universe. The wandering of the planets against the background of fixed stars on the sky is easy to observe, but only the most dedicated stargazers know firsthand the epicyclical path of the planets (Fig. 12.7). Epicycles can be drawn by combining two or more circular motions, as shown in Fig. 12.8 for Jupiter and Earth. If planetary orbits were truly circular, the Ptolemean approach would yield exactly what an earthbound observer sees.

The inverse problem, reconstructing true orbits from an "over the edge" view of their geocentric counterparts (as in Fig. 12.7), must have been a far more daunting task for Ptolemy to perform. Yet he mastered it accurately enough to reveal the discrepancies between the results of his methods and observations. They were caused—as we now know—by the orbits' ellipticity; trapped in the circular motion concept, he compensated for them by slightly offsetting the centers of revolution of the superimposed orbital

Figure 12.7   Three years of Jupiter's apparent orbit through the zodiac

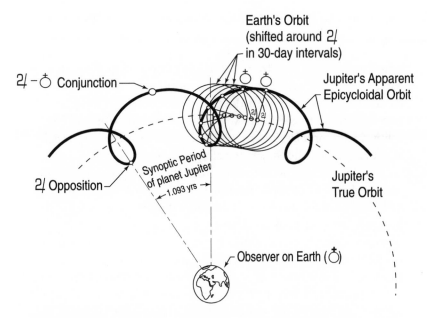

Figure 12.8 Superposition of the motions of Earth and Jupiter yields the epicyclical orbit that we actually see in the night sky.

circles. Ptolemy did not know he was just a step away from setting planets into their true, elliptical paths because the combination of two circular motions of equal periods but opposite sense (clockwise and counterclockwise) actually generates an ellipse (Fig. 12.9) as a special case of epicycloidal curves. But he missed what was to become the essence of Kepler's first law: positioning the sun at the focal point of the planets' elliptical orbits rather than at their center. This assumption must be seen as Kepler's epoch-making breakthrough, which allowed Isaac Newton to deduce the law of gravitational attraction.

Ptolemy describes *how* celestial bodies move, while Newton explains *why* they move that way.

No Units in Math . . .

That the presentation of the laws of nature can be done without the use of units shouldn't surprise us. After all, mathematics, the

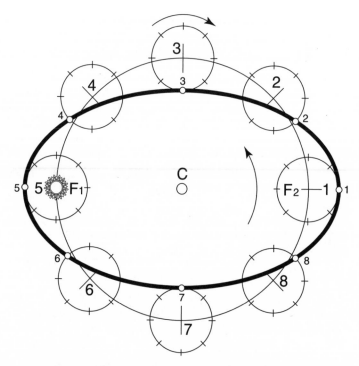

Figure 12.9 The ellipse as a special case of an epicycloid

science that inherently covers them all, is in itself dimensionless, which is the reason for the absolute precision of mathematical statements.

By way of example, Newton's law of gravity, setting the gravitational attraction of two masses m_1 and m_2 proportional to their product, $m_1 \times m_2$, presupposes that gravity, one of the forces of nature, complies with the purely mathematical operation of multiplying m_1 by m_2. From the four basic operations—addition, subtraction, multiplication, and division—to algebra, calculus, vector analysis, and complex function theory, the rules of mathematics never failed as the infrastructure for the formulations of the laws of nature. As Nobel Prize–winning physicist Eugene Wigner famously summed it up, there is an "unreasonable effectiveness of mathematics in the physical sciences."

If the Units Walked Out . . .

Often it helps our understanding of natural phenomena to forgo the use of units and express scientific data without dimensions. Defining the Moon's mass as $1/81$ the mass of Earth, or Jupiter's mass as $1/1045$ that of the Sun, conveys a far better picture of the size of those celestial bodies than stating their masses as 7.343×10^{22} and 1.900×10^{27} kg, respectively. Likewise, the fact that the oceans cover 0.708 (70.8%) of the Earth's surface is easy to visualize, while 148.8 million square kilometers or 57.45 million square miles are impossible to fathom.

The Nihilists and the Oddballs

William Froude, a British naval engineer, was the first to measure drag and sideforce on prototypes of ships in what became known as a "towing tank," a stagnant water canal that Froude constructed in 1870 at his house, Chelston Cross, in England.

Such setups are common in today's naval engineering world. The models under test are anchored over a dynamometer and tethered to a self-propelled overhead towing carriage. Drag readings from the dynamometer, complemented by visual inspection of liquid flow patterns, have since become basic tools in streamlining the shapes of vessels for lowest drag and highest efficiency.

The importance of Froude's discovery lies in his method of converting such results from scaled-down models to the "real thing," such as full-sized ocean liners. Since the principal forces that act on a ship in motion are gravitational and inertial, hydrodynamic flow patterns can be expected to be similar as long as the ratio of these forces is kept constant. This principle allows the use of prototypes to predict the performance of full size hulls.

With L the characteristic length dimension, such as a ship's or model's overall length or width, and g the free-fall acceleration (9.80665 m/sec^2), gravitational forces on a liquid element within the liquid flow pattern around the ship are proportional to $L \times g$.

Inertial forces, on the other hand, are known from the equations of hydrodynamics as proportional to the square of the ship's velocity, v, which led to the definition of "Froude's number":

$$\text{Fr} = \frac{\text{inertial force}}{\text{gravitational force}} = \frac{v^2}{L \times g}, \text{ which is dimensionless.}$$

With L_1 and v_1 for length and velocity of the prototype, and L_2 and v_2 for the actual vessel, the Froude numbers are $v_1^2/(L_1 \times g) = v_2^2/(L_2 \times g)$, from which we get

$$\left(\frac{v_2}{v_1}\right)^2 = \frac{L_2}{L_1}, \text{ or still } \frac{v_2}{v_1} = \sqrt{\frac{L_2}{L_1}}.$$

For instance, the velocity of a 1:100 scaled-down model must be $\sqrt{1/100} = 1/10$ of the ship's rated speed.

Should we, some day in the future, succeed in "terraforming" planet Mars, shipbuilders could still use earthbound measurements. With gravitation on Mars 0.38 of that on Earth, that is, $g_M = 0.38\, g$, Froude's numbers become $v_1^2/L_1 \times g = v_2^2/L_2 \times g_M$. For ships of the same length as on Earth, we have $L_1 = L_2$, and thus

$$\frac{v_1^2}{g} = \frac{v_2^2}{g_M} \text{ or } \frac{g_M}{g} = \left(\frac{v_2}{v_1}\right)^2 = 0.38 \text{ or still } \frac{v_2}{v_1} = \sqrt{0.38} = 0.616.$$

Ocean liners sailing the Martian seven (?) seas would have to reduce their cruising speed to 61.6% of that on Earth in order to maintain all other conditions equal.

Dimensionless transfer numbers also exist in aerodynamics, such as the widely known Reynolds number, $R = \rho \times v \times L/\eta$, with ρ for density and η for dynamic viscosity of air, originally proposed by G. G. Stokes in the 1850s. Model ships get hoisted through the tow tank, but aircraft and airfoil prototypes remain stationary in a wind tunnel while the surrounding air is blowing past. Special wind tunnels for advanced aircraft design are capable of propelling air for short periods of time at velocities reaching Mach 15, or fifteen times the speed of sound.

Dimensionless numbers for converting test results from prototypes are commonplace in our day. In a great variety of applications, several hundred have so far been developed.

Tailored Units of Measure

Notwithstanding all past and present efforts to establish a unified system of weights and measures, every profession seems set on maintaining its own particular units. Typesetters use the pica (1/6 inch) and point (1/72 inch). The type in this book is eleven points and the "leading," the distance between lines, is thirteen points.

Professional boxers are classified by a worldwide system of particular units, from flyweight (112 lb) to heavyweight (above 178 lb). Seafarers around the world would hate to give up the U.S. nautical mile (6076 feet or 1852 m). Food labels list calories instead of joules, and blood pressure is read in millimeters of mercury head instead of the metric pressure unit pascal. Diamond traders stick to their carat unit of weight (0.200 gram or 3.086 grains), whether or not their countries' established systems require otherwise.

Outdated units of measure persist in popping up, such as the strict limit of twelve stones for a rider's bodyweight, set by a Scottish horse trader who leased his ponies to tourists for a scenic ride. For those who don't know, one stone equals 14 pounds.

The U.S. gallon we use today still equals the fifteenth-century British "wine gallon," while the British use the imperial gallon, which is equivalent to 1.20095 of our gallons. In 1423, an English statute defined the hogshead as 63 gallons, and barrels were made to hold that amount of wine or ale. Champagne, for those of more refined tastes, still comes in magnums and jeroboams.

While most of such units mirror the human foible of "hanging on to the old ways," other sets of units were shaped to the users' convenience. Long before our times, when Eratosthenes of Kyrene attempted to measure the Earth's circumference, he based his calculations on the distance between the ancient Egyptian cities of Alexandria and what hence has become Aswan. He used a unit of distance that sounds outlandish to our ears: camel-days, a perfect example of what could be called a "convenience unit."

It served Eratosthenes well, because the distance a caravan could cover per day was pretty uniform and well known. And yet, it would vary with weather and highway conditions, and thus lacks the constancy inherent to all "real" units.

Nevertheless, using time instead of length to express distances is not entirely unheard of, even in our day. Ask a guide in the Swiss Alps how far it is to the top of a certain mountain, and you may hear something like: "Five hours." This way of expressing himself is more useful than if he had responded "four kilometers level and two thousand meters up." In Los Angeles, people give you distances in terms of "a half hour, without traffic," but in L.A. there is always traffic.

"Convenience units" pop up everywhere. For instance, there is the "slump factor" of concrete mixes, defined by the loss of height following the withdrawal of the mold from a freshly poured concrete slug. Initially it was molded as an inverted, truncated cone of specified dimensions, resembling the lower left of Figure 12.3.

Other examples are the many units expressing viscosity, the "internal friction" in liquids and gases. Viscosity is the principal factor in predicting the forces that transport fluids in pipelines, and it determines the liquid flow in spray painting, surface coating, and injection molding. Although its designated SI unit is the Pa·sec (pascal·second), industrial laboratories tend to express viscosity as the readouts of their instruments, mostly by the time, in seconds, for the fluid under test to empty a standard-size recipient through a calibrated orifice in its bottom. For thick, viscous liquids, like heavy oils and pitch, the falling sphere viscometer measures viscosity by the time a specific ball bearing takes to sink through a liquid column of a certain height—here again in seconds. A once often used unit, the centipoise (cP), equals 10^{-3} Pa·sec. Another gadget, the Engler's viscometer, measures how long a given amount of a fluid takes to pass through a calibrated capillary. This ratio of that time divided by the time for pure water is known as the dimensionless "Engler degree."

These "instrument-linked" units—convenient as they may be—were not derived from the basic units of any of the current systems of measures and therefore cannot be used "as is" in the evaluation of formulas. For instance, the unit of time cannot, all

of a sudden, be used for something entirely different, like the resistance of a liquid to flow. Saybolt seconds, Redwood seconds, and the like are not true units of viscosity, but a measure of viscosity. To use them with any of the equations of physics, they must be converted beforehand.

Take, for instance, Stokes' law for the viscous drag F_v on a ball of radius r, traveling with the velocity v through a fluid of dynamic viscosity η:

$$F_v = 6\pi \times \eta \times r \times v.$$

Any attempt to enter η in, say, "Redwood No. 1 seconds" or "Engler degrees" would of course give entirely erroneous results. Table 12.1 lists the kinematic viscosity of automotive commercial petroleum oils at 100 °F in units of the International System, along with the related Engler, Saybolt, and Redwood figures. Herein, kinematic viscosity stands for the ratio of dynamic viscosity, η, divided by the fluid's density, ρ, and carries the metric dimension of m^2/sec. Note that the figures in Table 12.1 are empirical values, meaning that we do not have conversion formulas. By contrast, conversion tables, like those in the appendix of this book, are entirely based on conversion formulas of the kind found in chapter 9.

The laws of physics can be shown in units of convenience when these seem more useful. For instance, the formula $s = v \times t$, for the distance s covered in the time t by a vehicle moving at the velocity v, yields s in meter if we input v in meters per second and t in seconds. But those who would rather get s in kilometers can use

Table 12.1. Comparison of Viscosity Measures

Type	Dynamic Viscosity η/ρ in 10^{-6} m^2/sec	Degrees Engler	Saybolt-sec (38 °C)	Redwood-sec
SAE 10W	41	5.46	190	167
SAE 20W	71	9.37	327	288
SAE 30	114	15.01	525	462
SAE 40	173	22.77	796	701
SAE 50	270	35.50	1243	1093

the relation $L(\text{km}) = 0.001 \times L(\text{m})$ and rewrite the equation into $s = 0.001 \times v \times t$. Or, for s in nautical miles, the relations 1 nautical mile = 1852 meters leads to $s = v \times t/1852$.

What makes this possible is that meter, kilometer, nautical miles, and so on are all dimensionally equal, representing length. By contrast, Stokes' equation cannot be rewritten for use with Engler degrees or Saybolt seconds.

In the microcosm of nuclear physics there is an unprecedented explosion of units. For length we have the fermi (fm), named after the Italian physicist Enrico Fermi, the first to succeed in generating a chain reaction in Chicago's nuclear reactor in 1942. Privately, Fermi was known for his belief in systematic procedure, which, incidentally, caused him to arrive late for his wedding. When he found that the sleeves of his shirt, selected to go with his tuxedo, were way too long for his arms, he quietly sat down at his mother's sewing machine and stitched folds into both sleeves.

Fermi and Franco Rasetti, his close friend from the days he was at the University of Rome, used to pester their friends and families by demanding, and also offering, scientific explanations of everyday phenomena. But their knowledge of U.S. history was limited. The story goes that on a trip down the East Coast, they came across a historical marker inscribed "The Mason-Dixon line." "It's a line along the Mason and Dixon rivers, dividing the North from the South," was Rasetti's guess.

Fermi, undeterred, replied, "Rivers? Mason and Dixon were American senators, one from the North, the other from the South."

Neither, of course, was correct. In pre-Civil War days, two Englishmen, Charles Mason and Jeremiah Dixon, surveyed the line along the 39°43' parallel as the common border of the land grants of the Penns and the Calverts, respective owners of Pennsylvania and Maryland. Mason and Dixon, unlike Fermi and Rasetti, used the foot and the chain.

One of Fermi's lesser-known creations, the remotely controlled "Fermi trolley," used to bring specimens to be irradiated into reach of a pick-and-place device at the University of Chicago's cyclotron. Driven by the machine's inherent magnetic field, the Fermi trolley would run all by itself around the top of

the cyclotron's vacuum box, freeing the operators from work in radiation-endangered areas. Fermi's trolley was the predecessor of the robots that we, the children of the computer age, take for granted as the devices that, among a number of other tasks, weld our cars' chassis and paint their bodies.

Fermi gave his name to the atomic length unit of one femtometer, or 10^{-15} m. The hydrogen nucleus measures 2.4 fermis, and the entire hydrogen atom is about 200,000 fermis across, so that an illustration of Bohr's atomic model cannot be drawn to scale on a sheet of paper of manageable size (Fig. 12.10).

Much like astronomers made the average distance from Sun to Earth an astronomical unit (AU), many nuclear physicists and chemists prefer to use "Bohr radii," the radius of the orbit of the electron in a hydrogen atom at its lowest energy state, rather than fermi. It measures $a_0 = 0.5292 \times 10^{-10}$ m.

The atom of the lightest of all elements, hydrogen, was the initial choice for the atomic scale unit of mass, of 1.6735×10^{-27} kg. But the hydrogen atom is unique among all elements insofar as its nucleus consists of a sole proton, while all other nuclei are a bond of protons and neutrons. Since a proton carries a positive electric charge, one would expect it to be somewhat more massive than the chargeless neutron, but the opposite is the case: the proton is about 0.14% less massive than the neutron.

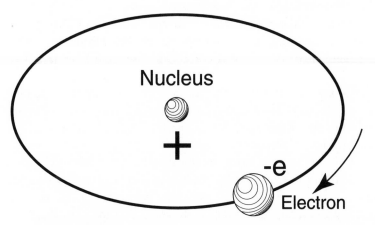

Figure 12.10   Niels Bohr's model of the hydrogen atom

This difference inspired the concept of an atomic mass unit based on the mass of both protons and neutrons. Carbon $^{12}_{6}$C, the principal ingredient of coal, and—by the way—the basic element of organic chemistry, was used instead of hydrogen in the definition of a unified atomic mass number (Fig. 12.11). The carbon nucleus consists of six protons and six neutrons, so that the new unit of mass equals $1/12$ of the nuclear mass of $^{12}_{6}$C.

This made the mass of the hydrogen nucleus slightly larger than the carbon-based unit mass, namely 1.007825 u. For comparison, the mass of the nucleus of weapon-grade uranium, consisting of 92 protons and 143 neutrons, is 235.043945 u.

Further on, the electron volt (eV) became the atomic scale unit of energy, although its name resembles electrical potential more than the counterpart of the macrocosmic energy unit of joule. Joule is, by the way, also the name of a crater, located at 27.3° N

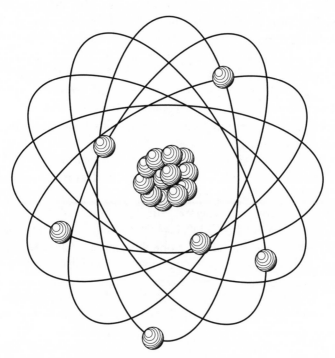

Figure 12.11   The carbon atom

and 144.2° W on the Moon. Both the unit of energy and the Moon's crater are named after James Prescott Joule, son of a nineteenth-century British beer brewer. The young Joule is best known for establishing the mechanical equivalent of thermal energy which, converted into metric units, is 1 kilocalorie = 427 meter·kilogram·force. Herein, the calorie is the amount of thermal energy causing a 1° Celsius increase in the temperature of 1 kg of water. The enormous amount of mechanical work for generating one single kilocalorie of heat must have come as a big surprise to Joule's contemporaries, in particular because the power output of James Watt's steam engines lagged far behind such figures. We now are aware that the conversion of mechanical work into heat is possible without losses, but that the inverse is not true, because we live in a kind of "preheated" environment. For instance, we tend to consider the heat content of one liter of boiling water as 100 calories, but that's relative to the arbitrary zero point of the Celsius temperature scale. The boiling water's absolute heat content must be figured from the absolute zero temperature point, −273 °C, which brings it up to 373 kcal. Now comes the twist: Can we recuperate those 373 kcal? Of course not. At best, we can get back what we invested to get the water boiling. Likewise, the heat of combustion of common gasoline, 11,500 kcal/kg, should drive our car a long way indeed. But for the reasons shown, only a small fraction of that energy can be converted into mechanical work.

Joule's name made it into the International System of Units for the unit of newton·meter for energy and work. On the other hand, the atomistic energy unit, the electron volt, is defined as the energy for moving an electron through a one-volt potential difference in an electric field. Take, for instance, a plate capacitor connected over a switch to a one-volt battery. Close the switch, and electrons sitting on the negative plate get propelled by electrostatic forces toward the positive plate, where they arrive with the velocity of v = 593,000 m/sec. With the mass of the electron, m_e = 9.1096 × 10^{-31} kg, the kinetic energy formula $E = mv^2/2$ gives for the kinetic energy of the electron:

$$1 \text{ eV} = \frac{1}{2} \times 9.1096 \times 10^{-31} \times 593,000^2 = 1.602 \times 10^{-19} \text{ joule.}$$

With that unit, the energy of the blue light photon of 440 nm of wavelength equals 2.82 eV, while mighty atom smashers, such as the Stanford Linear Accelerator (known as SLAC), operate in the range of giga (10^9) electron volts.

On the other end of the spectrum of length units, the distance between Sun and Earth, 149.5979×10^6 km or 92.9558×10^6 miles, was made the astronomical unit (AU) for conceptualizing the solar system. For instance, the 778.3×10^6 km distance between planet Jupiter and the Sun becomes 5.20 AU, easily visualized as "5.2 times farther from the sun than Earth is."

A light-year is the distance traveled in a year by a ray of light. In stellar and galactic surveys, the light-year is frequently used as a unit of length. Our closest stellar neighbor, apart from the Sun, is Alpha Centauri, some 4.34 light-years away. Its light reaches us earthlings 4.34 years after it is emitted. If we had a big enough telescope to discern on the star's surface something like "Alpha Centauri spots," counterparts to our sunspots, the ones we see today would have formed 4.34 years ago. Mostly, though, astronomers use parsecs, or kiloparsecs, rather than light-years.

An interesting attempt at the light-year's metrication was made in the 1930s by the Austrian astronomer Oswald Thomas, who introduced the "light-mile," equivalent to 10×10^{12} km. This integer unit, some 5 percent larger than the traditional light-year of 9.460550×10^2 km, would have been highly practical, but it fell victim to the light-year's strength of expressing distance and time lapse by the same number. Thus, the light-mile, though it is true to metric principles, never caught on.

Measurements Created Units

Unconventional units are often powerful tools for enhancing our mental picture of the world we live in. Measurements, too, are often easiest in appropriate, nonconventional units.

Stellar distances are found by a method similar to triangulation used by surveyors, which led to the unit of parsec. The word was formed by combining parallax and (arc)second and is based on the periodical shift in the apparent positions of nearby stars against the distant stellar background, a shift due to the Earth's motion

around the Sun. A star's parallax is thus its angular change of position relative to a one-AU displacement of Earth at right angles to the star's line of vision. One arcsecond of parallax would make for one parsec of distance, although no stellar object is located that close to the Sun. The parallax of 0.751 arcseconds for the nearest fixed star, Alpha Centauri, makes its distance $1/0.751 = 1.332$ parsec or, with the conversion factor of 3.26 light-years to the parsec, 4.34 light-years. Thus, parallax and stellar distance in parsecs are simply reciprocals of each other.

In 1610, Galileo Galilei made the earth-radius his unit of choice for measuring the distance to the Moon. Although most people find the Moon's face largest at moonrise, in reality our satellite's apparent size becomes greatest as it reaches its highest position on the sky. This is so because the Moon's distance from an earthbound observer (posted somewhere along the equator) lessens by one earth-radius as it wanders from the horizon toward the zenith (Fig. 12.12). Therefore, its apparent diameter, 30 minutes of arc (30') at the horizon, changes to 30.5' at the zenith— a 1/60 increase—indicating the moon's distance as sixty times the earth-radius (rounded down for the sake of simplicity—the exact

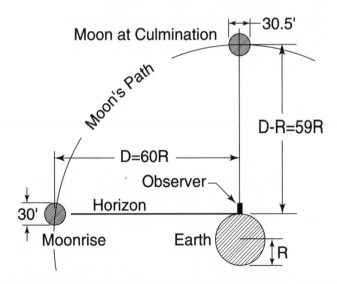

Figure 12.12 The Moon's distance derived from changes in its apparent diameter

ratio of the Moon's mean distance to the equatorial radius of Earth is 384,400 km/6378 km = 60.27).

In the early years of the heliocentric theory, astronomers could access only such relative data, since Kepler's laws allow for computing only the ratios of the planets' distances from the Sun with relation to one another. One way of linking these relative figures with earthbound units depends on the transit of the inner planet Venus over the Sun. In Figure 12.13 (not drawn to scale) an observer stationed on the North Pole and another on the South Pole would see the black disk of Venus pass at different solar latitudes, about 44 arcseconds apart. With the Earth-Venus distance known from Kepler's laws as 2/5 of the Earth-Sun distance, we use the law of similar triangles to find the equivalent of that angle as 5 earth-radii on the Sun's surface. Thus, the Sun-Earth distance, D, follows from $\tan 44'' = 5R/D$ as $D = 5R/\tan 44'' = 23439R$. With $R = 6357$ km for the polar radius of Earth, this allows us to estimate that $D = 23439 \times 6357 = 149 \times 10^6$ km.

Computing—In relative Figures: Artificial Satellites

The age of artificial satellites began in 1957 with the Russian *Sputnik*, followed by Wernher von Braun's launch of the United States' *Explorer I*.

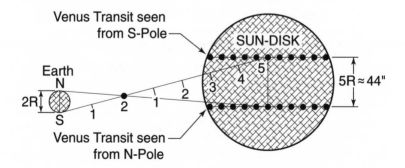

Figure 12.13 The distance Earth-Sun in Earth-radii, derived from observations of the transit of Venus from two or more different locations on Earth

While these early space vehicles circled the Earth in an hour and a half at about 1000 miles altitude, satellites of the Global Positioning System (GPS) take exactly 12 hours per turn. At Six groups of four each, a total of twenty-four, circle in six equally spaced circular orbits, of 26,560 km radius and 55° inclination relative to the equator.

The GPS receiver you might be carrying in your vest pocket contains a clock of sufficient precision to detect the time lag between release and reception of a characteristic signal from a satellite. Multiplied by c, the velocity of light and of electromagnetic waves in general, this interval equals the distance to the transmitting satellite, so that the transmissions from three satellites would suffice for triangulation of the location of the receiver and its owner on Earth. But that would be an open-ended measurement, prone to errors if the clock of the receiver were to fail to tick in perfect synchronicity with the clocks on the satellites. This can be avoided by adding one extra satellite, so that four simultaneous triangulations can be made. If correct, all should point out the same location; but if they fail to do so, a genial trick is used to correct location and timekeeping simultaneously: the receiver sets its clock automatically until all four triangulations agree with one another.

Those who lost their sleep over the Y2K virus could have worried about something else as the millennium drew to a close: GPS satellites started a particular binary calendar of 2^{10} weeks on January 6, 1980, and rolled over from 1023 to zero on August 22, 1999, restarting at 1980, as before. But except for the owners of certain outdated receivers, the switch passed unnoticed.

Still farther out in space are the geostationary communication satellites, which beam TV programs at domestic dish antennas; spy satellites orbit the Earth as well (Fig. 12.14). Their periods of revolution are synchronized with the rotation of Earth, which makes them hover over preselected geographical locations.

The period of revolution for geostationary satellites is 23 hours 56 minutes, which is less than the 24-hour interval between two successive culminations of the Sun. This is because the Earth rotates 366 times per year, which makes the period of its rotation, also called a *sidereal day*, 365/366 = 0.99727 average solar days.

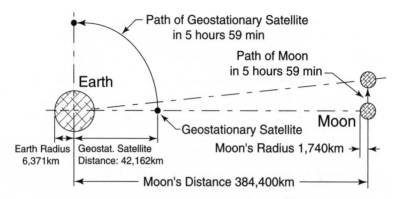

Figure 12.14 Earth, Moon, and geostationary satellite

With $24 \times 60 \times 60 = 86,400$ seconds to the day, the time a geostationary satellite must take to maintain its position in the sky, becomes $T = 0.99727 \times 86,400 = 86,164$ seconds. A satellite's orbiting altitude is computed from the condition that centrifugal forces F_c acting on the satellite's mass m_2 equal the tug of Earth's gravitational force F_g at the distance r of the satellite from the center of Earth: $F_g = F_c$. Gravitational force is given by Newton's law:

$$F_g = G \times \frac{m_1 \times m_2}{r^2}.$$

On the other hand, let a satellite of mass m_2, in circular orbit around Earth, take the time T per revolution. Centrifugal force F_c on that satellite then becomes

$$F_c = m_2 \times r \times \left(\frac{2\pi}{T} \right)^2,$$

where the term $2\pi/T$ is the satellite's angular velocity. Since gravitational and centrifugal forces cancel out, we have

$$G \times \frac{m_1 \times m_2}{r^2} = m_2 \times r \times \left(\frac{2\pi}{T} \right)^2.$$

Eliminating identical left and right side terms and rearranging the equation yields

$$r^3 = G \times m_1 \times \frac{T^2}{4\pi^2}.$$

This result contains Kepler's third law: the cube of a planet's distance from the Sun is proportional to the square of its time of revolution ($r^3 \propto T^2$). With $G = 6.670 \times 10^{-11}$ and $m_1 = 5.975 \times 10^{24}$ kg, we get for the distance of a geostationary satellite from the center of Earth:

$$r^3 = 6.670 \times 10^{-11} \times 5.975 \times 10^{24} \times \frac{86{,}164^2}{4\pi^2}$$

$$= 7.49473 \times 10^{22} \text{ and}$$

$$r = \sqrt[3]{7.49473 \times 10^{22}} = 42{,}161{,}744 \text{ m} = 41{,}162 \text{ km}.$$

With the Earth's equatorial radius of $R = 6377$ km, the satellite's altitude $h = r - R$ above the Earth's surface becomes

$$h = 41{,}162 - 6377 = 35{,}785 \text{ km} = 22{,}236 \text{ miles}.$$

The "Pick a Unit" Alternative

Instead of using data such as the constant of gravitation and the Earth's mass in kilogram, we can do with the familiar value of g for free-fall acceleration on Earth and R, the radius of Earth, to solve the same problem.

The weight of the satellite on Earth, $m_2 \times g$, equals the gravitational attraction between Earth and the satellite, $G \times m_1 \times m_2/R^2$, so that we have

$$m_2 \times g = G \times \frac{m_1 \times m_2}{R^2}, \text{ and so } G \times m_1 = g \times R^2.$$

This yields $G = gR^2/m_1$, and we have

$$G \frac{m_1 m_2}{r^2} = gR^2 \frac{m_2}{r_2} = m_2 r \left(\frac{2\pi}{T} \right)^2,$$

from which we deduce the simple-to-use term

$$r^3 = \frac{gR^2T^2}{4\pi^2}.$$

Finally, for a quick estimate, Kepler's third law, "the squares of the planets' sidereal periods are proportional to the cubes of their orbits' major semi-axes," can be used to deduce the orbital data of geostationary satellites from those of our most famous satellite, the Moon. However, the gravitational field of the Sun causes some alterations in the Moon's orbiting, making this method less precise than the former solutions. However, at times when neither the gravitational constant nor the mass of the Earth was yet known, this approach would have been our only alternative.

Kepler's third law, $(r_1/r_2)^3 = (T_1/T_2)^2$, used with Moon's period of revolution of T_2 = 27.34 days, the basis of the lunar month in the Jewish and Zulu calendars, and with T_1 = 0.99727 days for geostationary satellites, along with the Moon's distance of r_2 = 60.27 earth-radii, yields $(r/60.27)^3 = (0.99272/27.34)^2$, which gives for the distance $r = \sqrt[3]{60.27 \times 0.99727/27.34} = 6.629$ earth-radii or $6.629 \times 6367 = 42{,}280$ km. The slight difference between this and the previous results reminds us that Kepler's laws apply rigorously only to stand-alone gravitational systems of two bodies.

How Far Could We Get— Without Standard Units?

Although nature and its laws could be described by mere comparison of magnitudes, masses, and time periods, the resulting maze could become all but impenetrable. If you measure something relative to the Earth-radius and your peer somewhere else does it relative to the Earth-Moon distance, you both have a lot of work to do before you can compare answers. The knowledge would be there, but the complexities in sorting it out for the task at hand would grow exponentially with the quantity of data to be processed. That's why, instead of comparing natural phenomena directly with one another, we compare each with an *agreed upon set of standards* from a system of weights and measures.

· 13 ·

The Inter(galactic)net

Although a logically structured set of units, such as the International System of Weights and Measures, might be preferable, the use of arbitrarily selected units would in principle be possible as long as the laws of physics are formulated to accommodate them. Mother Nature has no preferences among all possible sets of units and leaves it to us humans to develop the ones best suited for our needs.

Even if we were to work with units defined merely by symbols rather than numerical values, we should succeed. For instance, the symbol λ could step in for meter or foot as the unit of length. μ could replace kgf or lb for the weight of the unit mass. And τ could become the unit of time. Used with the equations of physics, such units of measure should yield correct results, regardless of the numerical value attributed to each unit-symbol. λ could also stand for an inch, a centimeter, or say, 2.735 meters. Similarly, you may define μ as 0.379 kg, or if you will, 17.19 lb, and leave τ equal to the second, or make it an hour and a half—it shouldn't matter.

A Cry for Help from a Distant Civilization

Sounds farfetched, but it isn't pure theory. Envision a scenario where our radio telescopes, in search for signs of extraterrestrial civilizations, actually do receive signals from a planet circling a distant star. And further, that these signals, once deciphered, reveal a request for help in engineering some manageable project, say that of a hydroelectric power plant.

The planet would be too close to the mother star for us to see directly. Likewise, its inhabitants, even if they had powerful telescopes, could not get a glimpse of tiny planet Earth, hidden in the glare of the Sun. Like the Earth, the alien planet is not massive enough to cause a measurable wobble in the star's proper motion, which leaves us no way to work out its period of revolution. In short, all we know about the world from where the message came is the existence of a civilization anxious to upgrade its engineering capabilities with the help of ours. Our mission is to help them.

How to Build Dams

For starters, let's go over some basic technology necessary to build a dam.

The semi-schematic view of a hydroelectric plant (Fig. 13.1) shows dammed-up water from an artificial lake driving one or several turbines located at the lowest possible level at the foot of the dam. The total energy capacity will depend on the overall size of the barrage, but the *obtainable* power N_0 is proportional to the water head H, given by the difference in height from the liquid surface down to the turbines, and V, the amount of water per unit

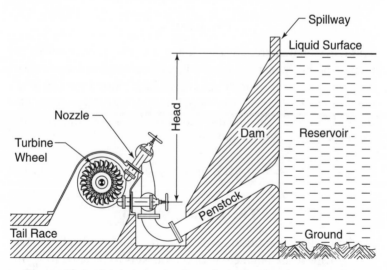

Figure 13.1 Semischematic view of a hydroelectric power plant

of time that can be consistently drained from the lake short of permanently lowering its level.

Energy Balance

With ρ for the density of the fluid powering the turbines—water on Earth, but some undefined liquid on the alien planet—the weight of the volume V of liquid flowing through the turbines per unit of time is

$$Q = V \times \rho. \tag{13.1}$$

In theory, this quantity of liquid, descending from the level H ("head"), would generate the power of N_0,

$$N_0 = Q \times H = VH\rho, \tag{13.2}$$

but energy-conversion losses call for the introduction of the factor η for the plant's overall efficiency ($\eta < 1$), so that the expected power output becomes

$$N = \eta \times N_0 = \eta \times VH\rho. \tag{13.3}$$

Hydraulics

The crucial parameter for the design and the dimensions of the turbine is the velocity c_0 of the liquid jet at the point of impact with the buckets installed circumferentially around the machine's rotor or wheel. This velocity would be $c_0 = \sqrt{2gH}$ if it weren't for frictional losses in the pipelines and nozzles, amounting to about 2%, which turns this formula into

$$c_0 \approx 0.98 \times \sqrt{2gH}. \tag{13.4}$$

The Bucket Wheel

Under no-load conditions, we expect the bucket wheel's tangential velocity at the pitch circle of the buckets to be close to c_0, the velocity of the liquid jet. At the other extreme, excessive load

would cause the machine to stall. This means that v_1, the tangential velocity of the turbine wheel under normal operating conditions, should be found somewhere in between zero and c_0. The best results for single-nozzle machines are obtained with $v_1 = 0.45 \times c_0$, and for twin-nozzle turbines, as in our case, with

$$v_2 = 0.45 \times \sqrt{2} \times c_0 = 0.636 \times c_0. \qquad (13.5)$$

The Continuity Equation

When a given volume V of an incompressible liquid flows through a pipe, its velocity will be greatest through the narrow sections of the piping, and lowest where the pipes are widest, because the volume of liquid per second is constant throughout. For the supply line of diameter D_p (Fig. 13.2), this volume of liquid flow per second is the product of the pipe's cross-sectional area $A_p = 0.25\pi \times D_p^2$, and the flow velocity c_p in that part, that is,

$$V = A_p \times c_p = 0.25\pi \times D_p^2 \times c_p. \qquad (13.6)$$

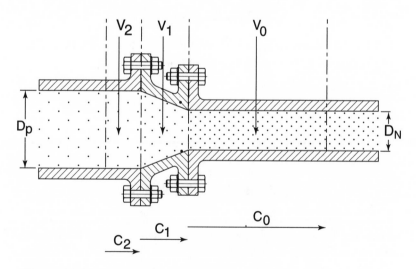

Figure 13.2 Liquid flow through pipes of varying cross section: the volumes of fluid in sections V_0, V_1, and V_2 are identical, while flow velocities increase inversely to the pipe's cross-sectional area.

That same volume of liquid flows through the narrower pipe of diameter D_N and area $A_N = 0.25\pi \times D_N^2$ at the (higher) velocity c_0, and is given by

$$V = A_N \times c_0 = 0.25\pi \times D_N^2 \times c_0. \tag{13.7}$$

Combined, these equations become $0.25\pi \times D_p^2 \times c_p = 0.25\pi \times D_N^2 \times c_0$, where 0.25π cancels out, which leads to the continuity equation of hydraulics:

$$\frac{D_p}{D_N} = \sqrt{\frac{c_0}{c_p}} \text{ or } \frac{c_p}{c_0} = \left(\frac{D_N}{D_p}\right)^2. \tag{13.8}$$

Rearranging Eq. (13.7) into $A_N = V/c_0$ and using the geometrical identity $A_N = 0.25\pi \times D_N^2$ gets us for the internal diameter at the exit of the nozzles:

$$D_N = \sqrt{\frac{4 \times V}{\pi \times c_0}}. \tag{13.9}$$

The Buckets

Bucket dimensions (Fig. 13.3) are linked to the exit diameter D_N of the nozzle by the empirical formulas:

$$L = 2.1 \times D_N \tag{13.10}$$

$$B = 2.5 \times D_N \tag{13.11}$$

$$T = 0.85 \times D_N. \tag{13.12}$$

Dimensions of the Bucket Wheel

For good balance between the sizes of hub and buckets, the pitch diameter D of the bucket wheel should amount to at least eight times the buckets' L-dimension (Eq. 13.10), that is,

$$D \geq 8 \times L. \tag{13.13}$$

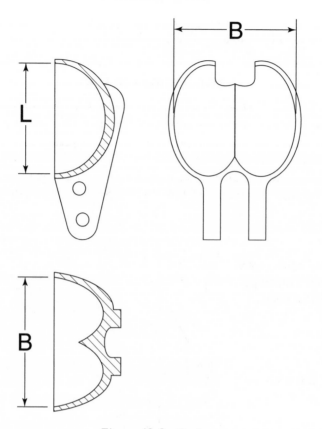

Figure 13.3   Buckets

At the pitch circle of diameter D and circumference $D \times \pi$, the tangential velocity of the bucket wheel when it rotates at n revolutions per unit of time is

$$v_2 = n \times D \times \pi, \text{ or inversely, } n = \frac{v_2}{D \times \pi},$$

$$\text{or still } D = \frac{v_2}{n \times \pi}. \tag{13.14}$$

This equation yields the number of rotations of the bucket wheel per unit of time. But let's keep in mind that our familiar rpm

accounts for revolutions per minute and needs to be divided by 60 for use in formulas such as the above.

Dimensioning the Dam

The height H of the dam is determined by the topography of the site, but its width b at the bottom (Fig. 13.4) is calculated to avoid the dam being toppled by the water pressure from the lakeside.

The dam, schematized in Figure 13.4 with a triangular cross section, has the volume of $0.5 \times b \times H$ per unit of length. With

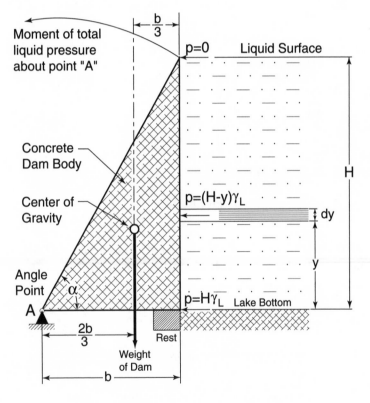

Figure 13.4   Statics of a concrete dam

ρ_c the density of the material the dam is made from (i.e., concrete), its weight per unit length becomes

$$W_C = 0.5 \times b \times H \times \rho_C. \qquad (13.15)$$

Since the center of gravity of a triangular area is located at one-third of the triangle's height, the leverage of the dam's weight around the tilt point A in Figure 13.4 is $2b/3$. Thus, the torque T, counteracting the water pressure, becomes

$$T = \frac{2}{3} \times W_C \times b, \qquad (13.16)$$

which with Eq. (13.15) leads to

$$T = \frac{2}{3} \times \frac{1}{2} \times b^2 \times H \times \rho_C = \frac{1}{3} \times b^2 \times H \times \rho_C. \qquad (13.17)$$

This torque keeps the dam from being overturned by the weight of the liquid in the reservoir. The countertorque, which tends to topple the dam, can be derived from ρ_L, the liquid's specific gravity, and the liquid pressure against the dam at the depth $(H - y)$ below the surface (Fig. 13.4):

$$p = \rho_L \times (H - y), \qquad (13.18)$$

which gives for the partial load dF on an area of unit length and the height of dy

$$dF = p \times dy = \rho_L \times (H - y) \times dy. \qquad (13.19)$$

As the force dF acts along a line distant by y from point A, the torque around A, exerted by dF, is

$$dT = dF \times y = \rho_L \times y \times (H - y) \times dy. \qquad (13.20)$$

Integration yields for the total torque T from the side of the lake:

$$T = \rho_L \times \int_0^H (H \times y - y^2) \times dy = \rho_L \times \left[\frac{H \times y^2}{2} - \frac{y^3}{3} \right]_0^H$$

$$= \rho_L \times \left(\frac{H^3}{2} - \frac{H^3}{3} \right), \text{ that is}$$

$$T = \frac{1}{6} \times \rho_L \times H^3. \tag{13.21}$$

This must equal the torque from Eq. (13.17), so that we get

$$\frac{1}{6} \times \rho_L \times H^3 = \frac{1}{3} \times \rho_C \times H \times b^2 \text{ and, canceling out,}$$

$$\rho_L \times H^2 = 2 \times \rho_C \times b^2.$$

From that equation follows $(b/H)^2 = \rho_L/2 \times \rho_C$ and the minimum ratio of width to height

$$\frac{b}{H} \geq \frac{1}{\sqrt{2}} \times \sqrt{\frac{\rho_L}{\rho_C}}. \tag{13.22}$$

Thus, the ratio of the dam's width b to its height H must be greater than the square root of the ratio of the density of the liquid in the lake (water on earth, but unknown on the alien planet), ρ_L, and the density of the material of which the dam is made, ρ_C, divided by $\sqrt{2}$.

Bridging the Interstellar Gap

With these formulas for the design of the power plant, we are prepared to solve the problem of our stellar neighbors, of whom we know very little beyond some physical constants in their particular units of measure, such as λ = unit of length, μ = unit of weight, and τ = unit of time. Let's thus assume that we received the following data from the alien planet:

Density of the liquid used for power generation: $\rho_L = 74\ \mu/\lambda^3$
Density of dam-building material: $\rho_c = 180\ \mu/\lambda^3$
Gravitational acceleration: $g = 15\ \lambda/\tau^2$
Liquid level above turbine location: $H = 1600\ \lambda$
Liquid flow per unit of time: $V = 60\ \lambda^3/\tau$

The Alien Power Potential

These data fail to make any sense to us Earthlings because they are in the undefined units of λ, μ, and τ, yet they allow us to specify the aliens' power plant. First, we get the weight of the descending liquid per unit of time from Eq. (13.1):

$$Q = V \times \rho_L = 74 \times 60 = 4440\ \mu/\tau,$$

and we find the available power from Eq. (13.2):

$$N_0 = Q \times H = 4440 \times 1500 = 6.66 \times 10^6\ \lambda\mu/\tau.$$

Assuming the plant's overall efficiency with $\eta = 0.80$ (80%), its effective power output becomes

$$N = \eta \times N_0 = 0.8 \times 6.66 \times 10^6 = 5.328 \times 10^6\ \lambda\mu/\tau.$$

Turbine Dimensioning

To design the turbine, start with computing the exit velocity of the liquid jet from the nozzles by Eq. (13.4):

$$c_0 = 0.98 \times \sqrt{2gH} = 0.98 \times \sqrt{2 \times 15 \times 1500} = 207.9\ \lambda/\tau.$$

The tangential velocity of the buckets follows from Eq. (13.5):

$$v_2 = 0.45 \times \sqrt{2} \times c_0 = 0.636 \times c_0 = 0.636 \times 207.9 = 132.2\ \lambda/\tau.$$

Before we use Eq. (13.14), $n = v_2/(D \times \pi)$, with that figure, we must get an estimate for the wheel diameter D from Eq. (13.13),

$D \geq 8 \times L$. However, D depends on the width L of the buckets, which again depends on D_N, the free nozzle diameter.

   In calculating the latter, we must remember that in a twin nozzle machine (Fig. 13.1), only half of the liquid volume V passes through each nozzle, so that the value of $60/2 = 30 \; \lambda^3/\tau$ is to be used for the liquid flow V. Thus, we get the diameter of the nozzles from Eq. (13.9):

$$D_N = \sqrt{\frac{4 \times V}{\pi \times c_0}} = \sqrt{\frac{4 \times 30}{207.9 \times \pi}} = 0.428 \; \lambda.$$

Hence, the dimensions of the buckets are deduced from

   Eq. (13.10): $L = 2.1 \times D_N = 2.1 \times 0.428 = 0.899 \; \lambda$

   Eq. (13.11): $B = 2.5 \times D_N = 2.5 \times 0.428 = 1.070 \; \lambda$

   Eq. (13.12): $T = 0.85 \times D_N = 0.85 \times 0.428 = 0.364 \; \lambda.$

The pitch diameter of the rotor, at least eight times the size of L, is now $D_N \geq 8 \times 0.899 \; \lambda$, that is, $D \geq 7.2 \; \lambda$. With the tangential velocity of the rotor given by Eq. (13.5) as $v_2 = 132.2 \; \lambda/\tau$, the turbine wheel's revolve is given by

Eq. (13.14): $\dfrac{v_2}{D \times \pi} = \dfrac{132.2}{7.2 \times \pi} = 5.84 \; \tau^{-1}$, and thus, $n \leq 5.84 \; \tau^{-1}$.

The closest integer, $n = 5 \; \tau^{-1}$, will be a good choice, so we set

$$D = \frac{v_2}{n \times \pi} = \frac{132.2}{5 \times \pi} = 8.416 \; \lambda.$$

And the Piping?

What's left is sizing the supply piping. According to a rule of thumb for keeping frictional losses low, the pipes should be wide enough to make the liquid flow velocity in the pipes no more than 13 percent of that through the nozzles, that is, $c_p = 0.13 \times c_0$. Thus we obtain from Eq. (13.8):

$$\frac{D_p}{D_N} = \sqrt{\frac{c_0}{c_p}} = \sqrt{\frac{c_0}{0.13 \times c_0}} = \sqrt{\frac{1}{0.13}} = 2.7.$$

With $D_N = 0.428\ \lambda$, this yields for the diameter of each nozzle's supply pipe $D_p = 2.77 \times 0.428 = 1.19\ \lambda$, and for the diameter D_{pp} of the penstock piping, sized for twice the rate of flow of the branch piping to the nozzles: $D_{pp} = 1.19 \times \sqrt{2} = 1.68\ \lambda$.

We Did it!

At this point, let us sit back and contemplate what we have done for our alien friends. If their data were correct, and if they carefully followed our instructions, their power plant will provide them with all the electricity obtainable from the natural resources within the scope of the data they gave us.

Of course, we shall never find out how much that really is. Neither can we have any idea of the size of the plant they might be erecting to our specifications. If the alien λ unit comes close to our meter in length, their project might turn out to be gigantic, like Brazil's Itaipú on the river Paraná. But let λ be more like the centimeter, and the alien power plant becomes the size of the royal palace in Jonathan Swift's accounts of Captain Gulliver's travel to Lilliput, the land of the midgets. That's how it has to be, because scientific reasoning does not allow for human bondage. Science helps us navigate the enigmas of nature, but will slap our face every time we try to compress it into human-made molds.

· 14 ·

Units, Physics, and Mathematics

Socrates (470–399 B.C.), the venerated Greek philosopher, never wrote down a single word of what he said. Yet his Socratic method, a way of winning an argument by leading the opponent into contradicting himself, has inspired a flood of philosophical treatises. The celebrated Oracle at Delphi labeled him "the wisest, most free, just, and prudent man in the world," but that didn't save him from a sentence to "death by hemlock" for dissenting from the views of his city's rulers.

Despite his distaste for written evidence, some of Socrates' statements, such as "While others live to eat, I eat to live" sound quite appropriate for modern times, and his allegation that all Athenians were liars has fanned endless discussions on the validity of strictly logical thought. In the case in point, Socrates, a native Athenian, would be a liar himself, and therefore must have lied about his fellow citizens, which makes us deduce that in reality, Athenians were truthful. In that case, however—the reasoning continues—Socrates too must have been a truthful fellow, and his statement that all Athenians were liars was true, after all. That makes him, along with the rest of Athenians, liars, and therefore . . . and so on, in a never-ending series of alternately contradictory conclusions—a true Catch 22.

Was Socrates a liar after all, or wasn't he? Or were the principles of deductive logic a sham?

None of the above. He who told the story is lying, because by the laws of logic, Socrates could not have made that statement: if Athenians, including the philosopher, speak the truth, then Socrates, truthful himself, could never have described them as liars. Or if indeed they were liars, he, a liar, too, would have lied about it and—once again—called them truthful. Thus, whatever the moral standing of antiquity's Athenians, the one and only logically possible comments Socrates could have made was calling them truthful—let them be liars or not. The Bible (Titus 1:11–13) already refers to a similar problem with the quotation "One of themselves, a prophet of their own, said, Cretans are always liars."

What we have here is not a flaw in the ways of logical thought, but a series of logical deductions based on an inadmissible precondition. It is a classical example of the less than classical saying "garbage in—garbage out," which we computer addicts know so well from daily experience.

Mathematics and the Laws of Physics

Socrates' legacy surfaces in mathematics, where any theorem that does not lead to self-contradicting conclusions is considered true. For instance, there is no direct proof for the rule of algebra that $(-a) \times (-a) = + a^2$, and yet it has been accepted because the assertion $(-a) \times (-a) = -a^2$ leads into mutually contradicting conclusions. As W. H. Auden once wrote, "Minus times minus is plus, the reason for this we need not discuss."

Just as mathematical theorems are inferred from a set of axioms—statements that are unproven yet assumed correct—many of the equations of physics can be obtained by logical deduction from the obvious. Basic laws, such as the relation between depth and water pressure, the interdependence of gas volume and gas pressure, Ohm's law, and the laws of motion in a straight line are all plausible and follow common sense. Deductions based on thought rather than observation pervaded Aristotle's teachings on philosophy and physics and remained the basis of scientific thinking throughout the Middle Ages. Experimental research by Galileo Galilei on the laws of free fall was still regarded

as heresy for deviating from Aristotelian dogma. When church-men refused to look through his telescope to see the moons of Jupiter, they were simply adhering to Aristotelian principles. But once the dikes had been broken, experiment won. Theories were made to fit experimental data, rather than the other way round.

Common sense and intuition run out of steam when it comes to describing quantum or relativistic physics. For instance, the idea of atomic nuclei consisting of more than one single proton would contradict Coulomb's law of mutual repulsion of equal electrical charges (Fig. 14.1). Therefore, atoms heavier than hydrogen should disintegrate spontaneously unless we arbitrarily intro-duce a new force of nature, fittingly called "strong interaction," a force unknown in the everyday environment and alien to intuitive understanding.

Electrons, commonly thought of as the carriers of electric cur-rents, should generate electromagnetic waves as they orbit the atomic nucleus, just like the high-frequency alternating currents that flow through radio and TV antennas. But if that were the case, electrons would radiate their kinetic energy away and rapidly spiral into the nucleus. Such apparent contradictions were re-solved by introducing a series of quantum energy levels, which have no parallel in macrocosmic physics. Radiation is only emit-ted (or absorbed) when an electron jumps from one level to another.

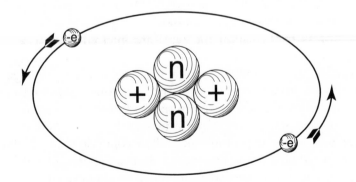

Figure 14.1 The helium atom:
n = neutron, + = proton, −e = electron

All too often, Mother Nature deviates from the expected, and so our theories must be consistently backed up by experimental data to make sure that the border between science and science fiction has not unwittingly been crossed.

Unlike mathematics, where the absence of contradictions helps to support a theorem, laws of physics require positive evidence. For instance, Einstein sought only static solutions for the universe in his general theory of relativity. He argued with the Russian scientist Alexander Friedmann, who said that there were solutions to Einstein's equations that suggested the universe was expanding. Both are logical; both follow from the equations. Only experimental evidence (such as data on the gravitational redshift of distant galaxies, counts of galactic radio sources, and detection of the cosmic microwave background radiation) can show which is right. Theoretical physics is akin to mathematics insofar as an entire system of equations can be derived from a few basic laws. A case in point is the derivation of the entire structure of celestial mechanics from Newton's law of gravitation and his laws of motion. Within their proper domain, such systems of equations are of absolute precision, like mathematical statements. But in the real world, they are destined to deviate to a certain degree from the results of pertinent observations or experiments, because the zillionfold interactions between matter and energy that constitute our universe are by far too complex for a comprehensive mathematical description.

Our basic "laws" of nature reside in an idealized, admittedly nonexistent environment. Physics textbooks treat the properties of a "perfect gas" and friction-free planes, and Kepler's laws apply to systems that consist of a central star with planets of negligible mass, situated in an entirely gravity-free environment save for the gravitation of the central star. But as we see in Figure 14.2, the center of mass of the solar system shifts and frequently resides outside of the Sun's body. The focal points of the orbits of the planets coincide with that center of mass, and therefore—regrettably—the real planets do not follow Kepler's predictions exactly. Admittedly, the law of gravitation allows for computing the "aberrations" from Kepler's elliptical orbits, caused by the mutual attraction of the planets and by the displacement of the center of

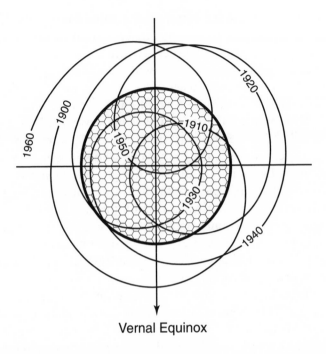

Vernal Equinox

Figure 14.2 Path of the center of mass of the solar system for the years 1900 through 1960

gravity of the solar system from the center of the Sun. But a host of other disturbances, such as the gravitational fields of a multitude of stars and dark matter in galactic space, are impossible to account for in their entirety.

Astronomers compute the onset of eclipses to fractions of a second and aim our planetary probes with awe-inspiring accuracy at such distant targets as Saturn's moons, yet none of their predictions will ever be absolutely precise. It is simply impossible to consider the interaction of all the trillions and trillions of masses throughout the universe, first because the task would be too much for any existing or imaginable supercomputer, and second because we do not have data on the vast majority of those bodies and probably don't even know about the existence of many of them.

Physical theories stand as monumental accomplishments in their own right but must be critically scrutinized to see how well

their predictions match the behavior of natural, experimental, or industrial setups. In the past, lack of uniformity of units of measure from country to country, and often even from county to county, made it difficult and sometimes impossible to compare the results of experiments done at a certain location with equivalent results obtained elsewhere.

The Price of Simplicity

Guidelines for the International System of Weights and Measures (SI) demand unit value for as many as possible of the proportionality constants in the formulations of the laws of physics.

The International System's use of 10-based subdivisions of units makes it attractive, uniform, and easy to use in all kind of calculations. That the spacing of the millimeter marks on a metric ruler helps distinguish them without eyestrain, yet leaves them close enough for good drawing precision, is an additional bonus, in particular for designers and draftspeople.

Last but not least, the equality of the mass of one liter of water with the kilogram, along with the identity of liter and cubic-decimeter, simplifies computations and avoids possible conversion errors between units and their subdivisions, which is common in other systems of measures.

Although the precision of the kg/dm^3 relation has been flawed by updated measurements that found the volume of one kilogram of water at the temperature of its greatest density (4 °C) to be 1.000028 liter, such subtleties should be the least of our metric conversion worries.

How Arduous Can It Get?

Most scientists have gone metric quietly, and the engineering community, having survived the switch from pen, pencil, and drawing board to mouse, keyboard and monitor, should have no trouble in adapting to metric units. With worldwide tourism within the reach of most pocketbooks, "we the people" have developed a great capacity for understanding strange units. One or two days abroad usually suffice to teach tourists, when not on the beach, how to haggle with local merchants in the currencies

of their host countries. So it is hard to understand why these same people, who effortlessly learned the conversion of the pound sterling, the yen, the drachma, and the yuan, seem outraged by the prospect of kilometer highway signs and kilogram scales.

Likewise, the venerable BTU was tearlessly abandoned by those watching their weight and even by those not watching it, to be replaced by the metrically based calorie. On the other hand, who has ever bought an engagement ring and asked for the "forever" diamond's weight in grains rather than carats?

The Global Vision: Illusion or Reality?

With taxpayers weary of paying more dollars, and the government set on balancing the budget rather than helping industry bear the costs of metrication, the chance that America will adopt the International System of Weights and Measures seems somewhat remote. Nevertheless, the vision of a worldwide unified system of units is appealing and, in principle, should rank higher than money pinching. But unfortunately, the cardinal question: "Can we do it?" has changed to: "What will it cost to do it?" and with that, our entire civilization has turned a corner.

When President John F. Kennedy asked whether we could beat the Soviets to the Moon, the answer was "yes" and we went. Since then, the question "Can we go to Mars?" has changed to "What's the cost of going to Mars?" and that's what keeps our astronauts in low orbit around Earth.

When the building of the superconducting supercollider at Waxahachie, Texas, was called off, it was because of the money. Senate and Congress viewed it more as pork-barrel spending than a fundamental contribution to basic science.

Thus, our metric conversion problems boil down to whether we have the stamina and courage to follow in the traditions of twentieth-century American technological achievements. If some of this pioneering spirit is still with us, a worldwide system of weights and measures will someday take foot quite naturally and without much ado.

Will we ever go the whole nine yards? Perhaps not, but we might go 8.23 meters!

· Appendix ·

Conversion Tables

Table A.1. The Meter Family of Length Units: Units of the International System.

	picometer pm	angstrom Å	nanometer nm	micrometer μm	millimeter mm	centimeter cm	decimeter dm	meter m	hectometer hm	kilometer km
1 picometer (fermi) =	1	0.01	0.001							
1 angstrom =	100	1	0.1							
1 nanometer =	1000	10	1	0.001	10^{-6}	10^{-7}	10^{-8}	10^{-9}		
1 micrometer =	10^6		1000	1	0.001	10^{-4}	10^{-5}	10^{-6}		
1 millimeter =			10^6	1000	1	0.1	0.01	0.001		
1 centimeter =			10^7	10^4	10	1	0.1	0.01		
1 decimeter =			10^8	10^5	100	10	1	0.1		
1 meter =			10^9	10^6	1000	100	10	1	0.01	0.001
1 hectometer =								100	1	0.1
1 kilometer =								1000	10	1

Other Metric Length Units

	m	km	AU	LY	pc
1 AU (astronomical unit) =	149.5979×10^9	149.35979×10^6	1		
1 LY (light-year) =	9.460528×10^{15}	9.460528×10^{12}	63.023973×10^3	1	0.3065948
1 pc (parsec) =	30.85678×10^{15}	30.85678×10^{12}	206.2648×10^3	3.261633	1
1 p (typographical point) =	127/360000	0.3527778×10^{-6}			

Notes: Cells left blank would show conversion factors of little practical value. One meter is defined as the 299,792,458th part of the path traveled by light in vacuum in the time interval of one second

Table A.2. The Foot Family of Length Units: Units of the U.S. customary System and their metric counterparts.

	league	nautical mile (US nat)	US statute mile (mi)	furlong	rod	yard (yd)	foot (ft) (form.: ')	inch (in) (form.: ")	mil (0.001 inch)	microinch (μinch)	kilometer (km)	meter (m)	centimeter (cm)	millimeter (mm)
1 league =	1	2.6069287	3	24	960	5280	15840	190080			4.828032	4828.032		
1 international nautical mile =	0.38359315	1	1.1507794	9.2062356	368.24942	2025.3718	6076.1155	72913.386			1.852	1852		
1 US statute mile =	1/3	0.86897624	1	8	320	1760	5280	63360			1.609344	1609.344		
1 furlong =	1/24	0.10862203	1/8	1	40	220	660	7920			0.201168	201.168		
1 rod =			1/320	1/40	1	5.5	16.5	198				5.0292		
1 yd =				1/220	2/11	1	3	36				0.9144	91.44	
1 ft =				1/660	2/33	1/3	1	12				0.3048	30.48	
1 inch =					1/198	1/36	1/12	1	1000	10^6		0.0254	2.54	25.4
1 mil =								0.001	1	1000		25.4×10^{-6}	25.4×10^{-4}	25.4×10^{-3}
1 microinch=								10^{-6}	0.001	1				25.4×10^{-6}
1 km =	0.20712373	0.53995680	0.62137119	4.9709695	198.83878	1093.6133	3280.8399	39370.079			1	1000	10^5	10^6
1 m =					0.19883878	1.0936133	3.2808399	39.370079			0.001	1	100	1000
1 cm =						0.010936133	0.03280840	0.39370079			10^{-5}	0.01	1	10
1 mm =						0.0010936133	0.0032808840	0.03937007	39.370079	39370.079	10^{-6}	0.001	0.1	1

Note: Cells left blank would show conversion factors of little practical merit. Conversion figures in shaded cells are exact.

Table A.3. The Kilogram Family of Mass Units: Units of the International System and their conversion to FPS units.

	metric ton t	kilogram kg	gram g	carat c	milligram mg	short ton short t	long ton long t	hdrd.wt. cwt	pound lb avdp	ounce oz avdp	dram dr avdp	grain gr
1 metric ton =	1	1000	10^6			1.1023113	0.98420653	22.046226	2204.6226			
1 kilogram =	0.001	1	1000	5000	10^6	0.0011023113	$0.98420653 \times 10^{-3}$	0.022046226	2.2046226	35.273962	564.38339	15432.358
1 gram =	10^{-6}	0.001	1	5	1000				0.0022046226	0.03527396	0.56438339	15.432358
1 carat =		0.0002	0.2	1	200					0.00705479	0.11287668	3.0864717
1 milligram =		10^{-6}	1/1000	1/200	1							0.015432358
1 short ton (FPS) =	0.90718474	907.18474				1	0.89285714	20	2000	32000		
1 long ton (FPS) =	1.01604691	1016.04691				1.12	1	22.4	2240	35840		
1 short hundredweight =	0.045359237	45.359237	45359.237			1/20	1/22.4	1	100	1600		
1 pound =		0.45359237	453.59237			1/2000	1/2240	0.01	1	16	256	7000
1 ounce =		0.028349523	28.349523	141.74762	28349.523	1/32000	1/35840	0.000625	0.0625	1	16	437.5
1 dram =		0.0017718452	1.7718452	8.8592260	1771.8452				1/256	1/16	1	27.34375
1 grain =		64.79891×10^{-6}	0.06479891	0.32399455	64.79891				1/7000	10/4375	1/27.34375	1

Note: Cells left blank would show conversion factors of little practical merit. Conversion figures in shaded cells are exact.

Table A.4. Units of Time: Units of the International System.

	year y	day d	hour h	minute min	seconds s	millisecond ms	microsecond µs	nanosecond ns
1 Mean Solar Year =	1	365.24219215	8765.8126116	525948.7567	31.55692540×10^6			
1 Mean Solar Day =	$2.7379093147 \times 10^{-3}$	1	24	1440	86400			
1 hour =	$114.07955478 \times 10^{-6}$	1/24	1	60	3600			
1 minute =	$1.9013259130 \times 10^{-6}$	1/1440	1/60	1	60	60000		
1 second =	$31.688765216 \times 10^{-9}$	1/86400	1/3600	1/60	1	1000	10^6	10^9
1 millisecond =					0.001	1	1000	10^6
1 microsecond =					10^{-6}	0.001	1	1000
1 nanosecond =					10^{-9}	10^{-6}	0.001	1

The various definitions of the Year and the Day

	Mean Solar Days d	Hours h	Seconds s
mean solar (tropical- or calendar) year =	365.24219215	8765.8126116	31.5569254×10^6
sidereal year =	365.25636	8766.15264	31.5581495×10^6
anomalistic year =	365.25964	8766.23136	31.5584329×10^6
common year (DIN 1301) =	365	8760	31.536×10^6
banker's year (interests' computing) =	360	8640	31.104×10^6
mean solar day =	1	24	86400
sidereal day (star day) =	0.997270	23.93447	86164.09

Notes: Cells left blank would show conversion factors of little practical merit. Conversion figures in shaded cells are exact.

Table A.5. Units of Force: The Newton family of units of force and its counterparts.

	newton N	poundal pd	kilogram-force kgf	pound-force lbf	ton-force tf	gram-force gf	milligram force (mgf)
1 newton =	1	7.2330139	1/9.80665	0.22480894	$0.10197162 \times 10^{-3}$	101.97162	101971.62
1 poundal =	0.13825495	1	0.014098082	0.031080950	14.098082×10^{-6}	14.098082	14098.082
1 kilogram-force =	9.80665	70.931635	1	2.2046226	0.001	1000	10^6
1 pound-force =	4.4482216	32.174049	0.45359237	1	$0.45359237 \times 10^{-3}$	453.59237	453592.37
1 ton-force =	9806.65	70931.635	1000	2204.6226	1	10^6	10^9
1 gram-force =	9.80665×10^{-3}	0.070931635	0.001	2.2046226×10^{-3}	10^{-6}	1	1000
1 milligram-force =	9.80665×10^{-6}	70.931635×10^{-6}	10^{-6}	2.2046226×10^{-6}	10^{-9}	0.001	1

Multiples and Fractions of the Newton

	meganewton MN	kilonewton kN	newton N	millinewton mN	micronewton μN
1 meganewton =	1	1000	10^6	10^9	10^{12}
1 kilonewton =	0.001	1	1000	10^6	10^9
1 newton =	10^{-6}	10^{-3}	1	1000	10^6
1 millinewton =	10^{-9}	10^{-6}	10^{-3}	1	10^3
1 micronewton =	10^{-12}	10^{-9}	10^{-6}	10^{-3}	1

Note: Shaded figures are exact.

Table A.6. Pressure Units: Units of the International System and their conversion to FPS units.

	pascal Pa	bar	kilogram-force per square-meter kgf/m²	technical atmosphere at	physical atmosphere atm	meter of water head m	Torr mm Hg	pound-force per sq.inch psi
1 pascal = 1 newton/square-meter =	1	10^{-5}	0.10197162	10.197162×10^{-6}	9.8692327×10^{-6}	101.97162×10^{-6}	7.5006376×10^{-3}	$0.14503773 \times 10^{-3}$
1 bar =	10^5	1	10197.162	1.0197162	0.98692327	10.197162	750.06376	14.503773
1 kilogram-force per square-meter =	9.806650	98.0665×10^{-6}	1	10^{-4}	96.784111×10^{-6}	10^{-3}	73.556127×10^{-3}	1.4223342×10^{-3}
1 technical atmosphere =	98066.50	0.980665	10000	1	0.96784111	10	735.56127	14.223342
1 physical atmosphere =	101325	1.01325	10332.275	1.0332275	1	10.332275	760.0021	14.695948
1 meter of water head =	9806.650	0.0980665	1000	0.1	0.096784111	1	73.556127	1.4223342
1 Torr = 1 mm Hg-column =	133.32200	1.3332200×10^{-3}	13.595060	1.3595060×10^{-3}	1.3157858×10^{-3}	13.595060×10^{-3}	1	0.019336720
1 pound-force per square-inch =	6894.7577	0.068947577	703.06962	0.070306962	0.068045968	0.70306962	51.715079	1

Note: Conversion figures in shaded cells are exact. International System figures boxed.

Table A.7. Units of Energy and Work: Units of the International System and their conversion to FPS units.

	joule J	kilojoule kJ	kilowatt-hour $KW{\cdot}h$	kilogram-meter $kgf{\cdot}m$	foot-pound $ft{\cdot}lbf$	horsepower-hour $hp{\cdot}h$ (MKS)	horsepower-hour $hp{\cdot}h$ (FPS)	kilocalorie $kcal$	British thermal unit Btu
1 joule =	1	0.001	$(5/18) \times 10^{-6}$	0.10197162	0.73756215	$0.37767267 \times 10^{-6}$	$0.37250614 \times 10^{-6}$	$10/41868$	0.9478134×10^{-3}
1 kilojoule =	1000	1	$(5/18) \times 10^{-3}$	101.97162	737.56215	$0.37767267 \times 10^{-3}$	$0.37250614 \times 10^{-3}$	$5000/20934$	0.94781337
1 kilowatt · hour =	3.6×10^{6}	3600	1	367.09784×10^{3}	2.6552237×10^{6}	1.3596216	1.3410221	859.84523	3412.1281
1 kilogram · meter =	9.806650	9.80665×10^{-3}	2.7240694×10^{-6}	1	7.2330139	$(1/270) \times 10^{-3}$	3.6530373×10^{-6}	2.3422781×10^{-3}	9.2948739×10^{-3}
1 foot · pound =	1.3558179	1.3558179×10^{-3}	$0.37661610 \times 10^{-6}$	0.13825495	1	$0.51205539 \times 10^{-6}$	$(1/1980) \times 10^{-3}$	$0.32383155 \times 10^{-3}$	1.2850624×10^{-3}
1 horsepower · hour (metric) =	2.6477955×10^{6}	2647.7955	0.73549875	270×10^{3}	1.9529137×10^{6}	1	0.98632007	632.41509	2509.6160
1 horsepower · hour (FPS) =	2.6845195×10^{6}	2684.5195	0.74569987	273.74481×10^{3}	1.98×10^{6}	1.0138697	1	641.18648	2544.4235
1 kilocalorie (SI) =	4186.8	4.1868	1.163×10^{-3}	426.93478	3088.0252	1.5812399×10^{-3}	1.5596087×10^{-3}	1	3.968305
1 British thermal unit =	1055.0600	1.0550600	$0.29307223 \times 10^{-3}$	107.58618	778.17234	$0.39846734 \times 10^{-3}$	$0.39301634 \times 10^{-3}$	0.25199676	1

Notes: Conversion figures in shaded cells are exact. International System figures boxed. Thermal Energy related figures based on the *International Table Calorie* of 4.186.8 joule, rather than the *Thermochemical Calorie* of 4184 joule.

Table A.8. Units of Power: Units of the International System and their conversion to FPS units.

	watt W	kilowatt kW	horsepower hp (metric)	horsepower hp (FPS)	meter·kilogram-force per second m·kgf/sec	foot·pound-force per sec. ft·lb/sec	kilocalorie (SI) per second kcal/sec	kilocalorie (SI) per hour kcal/h	British thermal unit per hour Btu/h
1 watt =	1	0.001	1.3596216×10^{-3}	1.3410221×10^{-3}	0.1019716213	0.73756215	$0.23884590 \times 10^{-3}$	0.85984523	3.4121281
1 kilowatt =	1000	1	1.3596216	1.3410221	101.9716213	737.56215	0.23884590	859.84523	3412.1281
1 horsepower (metric) =	735.49875	0.73549875	1	0.98632007	75	542.47604	0.17567086	632.41509	2509.6160
1 horsepower (FPS) =	745.69987	0.74569987	1.013869665	1	76.040225	550	0.17810735	641.18648	2544.4235
1 meter · kilogram-force per second =	9.80665	9.80665×10^{-3}	1/75	13.150934×10^{-3}	1	7.2330139	2.3422781×10^{-3}	8.4322012	33.461546
1 foot · pound-force per second =	1.3558179948	1.3558179×10^{-3}	1.8433994×10^{-3}	1/550	0.13825495	1	$0.32383155 \times 10^{-3}$	1.1657936	4.6262245
1 kilocalorie (SI) per second =	4186.8	4.1868	5.6924638	5.6145913	426.93478	3088.0252	1	3600	14.285898×10^{3}
1 kilocalorie (SI) per hour =	1.163	1.163×10^{-3}	1.5812399×10^{-3}	1.5596087×10^{-3}	0.11859300	0.85778478	1/3600	1	3.9683050
1 British thermal unit per hour =	0.29307223	$0.29307223 \times 10^{-3}$	$0.39846734 \times 10^{-3}$	$0.39301634 \times 10^{-3}$	29.885051×10^{-3}	0.21615898	69.999100×10^{-6}	0.25199676	1

Notes: Conversion figures in shaded cells are exact. International System figures boxed. Thermal Energy related figures based on the *International Table Calorie* of 4.186.8 joule, rather than on the *Thermo-chemical Calorie* of 4184 joule.

· Suggested Reading ·

Chapter 1: Blessing Our Countings

Hopp, Peter M. *Slide Rules: Their History, Models, and Makers.* Mendham: Astragal Press, 1999.

Kurtz, Max. *Handbook of Applied Mathematics for Engineeers and Scientists.* New York: McGraw-Hill, 1991.

Maor, Eli. *e: The Story of a Number.* Princeton, N.J.: Princeton University Press, 1994.

Turner, Gerald L'E. *Scientific Instruments, 1500–1900: An Introduction.* Berkeley: University of California Press, 1998.

Zebrowski, Jr., Ernest. *A History of the Circle: Mathematical Reasoning and the Physical Universe.* New Brunswick, N.J.: Rutgers University Press, 1999.

Chapter 2: Going to Great Lengths

Hightower, Paul. *Galilei: Astronomer and Physicist.* Hillside, N.J.: Enslow, 1997.

Chapter 3: Degrees of Separation

Casson, Lionel. *Ships and Seamanship in the Ancient World.* Baltimore: Johns Hopkins University Press, 1995.

Patai, Raphael, James Hornell, and John M. Lundquist. *The Children of Noah.* Princeton, N.J.: Princeton University Press, 1998.

Pérez-Mallaína, Pablo E. *Spain's Men of the Sea: Daily Life on the Indies Fleets in the Sixteenth Century.* Trans. Carla Rahn Phillips. Baltimore: Johns Hopkins University Press, 1998.

Sobel, Dava. *Longitude: The True Story of a Lone Genius Who Solved the Greatest Scientific Problem of His Time.* New York: Penguin, 1996.

Sullivan, Michael. *Algebra and Geometry.* New York: Prentice Hall, 1996.

Chapter 4: The "Obvious" Unit of Time

Arfken, George B., et al. *University Physics.* New York: Academic Press, 1984.

Einstein, Albert. *Relativity: The Special and the General Theory.* New York: Crown, 1995.

Mermin, N. David. *Space and Time in Special Relativity.* New York: McGraw-Hill, 1968.

Nahin, Paul J. *Imaginary Tale: The Story of* $\sqrt{-1}$. Princeton, N.J.: Princeton University Press, 1998.

Shankland, R. S. "The Michelson-Morley Experiment." *Scientific American* 211, 107 (November 1967).

Chapter 5: Weighty Matters

Haeder, W., and E. Gärtner. *Die gesetzlichen Einheiten in der Technik.* Berlin: Deutsches Institut für Normung e.V., 1980.

"Specifications, Tolerances, and Other Technical Requirements for Weighing and Measuring Devices." In *NIST Handbook* 44, Appendix C. Gaithersburg, Md.: National Institute of Standards and Technology, 1998.

Chapter 6: Gravimetric Standards

"Metric System of Measurement: Interpretation of the International System of Units for the United States." In *NIST Handbook* 46, no. 144. Gaithersburg, Md.: National Institute of Standards and Technology, 1998.

Chapter 7: The Matter with Mass

De Gandt, François, ed. *Force and Geometry in Newton's Principia.* Princeton, N.J.: Princeton University Press, 1995.

Chapter 8: Empire of Light

Mertz, Lawrence. *Excursions in Astronomical Optics.* New York: Springer-Verlag, 1996.

Weiss, Richard J. *A Brief History of Light and Those That Lit the Way.* New York: World Scientific, 1996.

Chapter 9: Hot Stuff

Cengel, Yunus A., and Michael A. Boles. *Thermodynamics: An Engineering Approach.* New York: McGraw-Hill, 1998.

Jones, J. B., and R. E. Dugan. *Engineering Thermodynamics.* New York: Prentice Hall, 1995.

Knowles Middleton, W. E. *History of the Barometer.* Baltimore: Johns Hopkins University Press, 1964.

———. *History of the Thermometer.* Baltimore: Johns Hopkins University Press, 1966.

Partington, J. R. *A History of Greek Fire and Gunpowder.* Baltimore: Johns Hopkins University Press, 1998.

Chapter 10: The Missing Link

Caneva, Keneth L. *Robert Mayer and the Conservation of Energy.* Princeton, N.J.: Princeton University Press, 1992.

Freidel, Robert, Paul Israel, and Bernard S. Finn. *Edison's Electric Light: Biography of an Invention.* New Brunswick, N.J.: Rutgers University Press, 1987.

Hebra, Alexius J. *Controlador Electrónico de Temperatura.* São Paulo, Brazil: Revista Rádio e Televisão, Monitor Institute, 1971.

Hogan, Brian J., and Alexius J. Hebra. *Unit Converts Windpower System's Output to 60 Hz Three-Phase Power.* Boston: Design News/ 2-6-84, Cahners Publishing Company, 1984.

Hughes, Thomas Parker. *Networks of Power: Electrification in Western Society, 1880–1930.* Baltimore: Johns Hopkins University Press, 1983.

Pera, M. *The Ambiguous Frog: The Galvani-Volta Controversy on Animal Electricity.* Trans. J. Mandelbaum. Princeton, N.J.: Princeton University Press, 1992.

Chapter 11: Compound Units

Bass, Hyman. *Measuring What Counts: A Conceptual Guide to Mathematics Assessment.* Washington, D.C.: National Academy Press, 1993.

Oberg, Erik, and Christopher J. McCauley. *Machinery Handbook.* New York: Industrial Press, 1996.

Chapter 12: Invasion of Aliens and Nihilists

Cooper, Dan, Enrico Fermi, and Owen Gingerich. *Enrico Fermi and the Revolutions of Modern Physics.* Oxford: Oxford University Press, 1998.

Copernicus, Nicholas, ed. *On the Revolutions.* Translation and commentary by Edward Rosen. Baltimore: Johns Hopkins University Press, 1992.

Morgan, Walter L., and Gary D. Gordon. *Communication Satellite Handbook.* New York: John Wiley and Sons, 1989.

Newton, Robert R. *Ancient Astronomical Observations and the Accelerations of the Earth and Moon.* Baltimore: Johns Hopkins University Press, 1970.

Newton, Robert R. *Ancient Planetary Observations and the Validity of Ephemeris Time.* Baltimore: Johns Hopkins University Press, 1976.

Newton, Robert R. *The Crime of Claudius Ptolemy*. Baltimore: Johns Hopkins University Press, 1977.

Raymo, Chet. *365 Starry Nights*. New York: Prentice Hall, 1982.

Stephenson, Bruce. *Kepler's Physical Astronomy*. Princeton, N.J.: Princeton University Press, 1994.

Toomer, G. J. *Ptolemy's Almagest*. Princeton, N.J.: Princeton University Press, 1998.

Wagner, Francis. *Eugene P. Wigner: An Architect of the Atomic Age. Highlights of a Career with a Comprehensive Bibliography*. Lanham, Md.: University Press of America, 1987.

Chapter 13: The Inter(galactic)net

Beyer, William H. *Handbook of Mathematical Sciences*. Akron, Ohio: CRC Press, 1987.

Magnusson, Roberta J. *Water Technology in the Middle Ages: Cities, Monasteries, and Waterworks after the Roman Empire*. Johns Hopkins Studies in the History of Technology. Baltimore: Johns Hopkins University Press, 2001.

Chapter 14: Units, Physics, and Mathematics

Baird, D. C. *Experimentation: An Introduction to Measurement Theory and Experiment Design*. New York: Prentice Hall, 1995.

Fitch, Val L., Daniel L. Marlow, and Margit Dementi. *Critical Problems in Physics*. Princeton, N.J.: Princeton University Press, 1997.

McClelland, James E., and Harold Dorn. *Science and Technology in World History*. Baltimore: Johns Hopkins University Press, 1999.

Thagard, Paul. *Conceptual Revolutions*. Princeton, N.J.: Princeton University Press, 1992.

· Bibliography ·

Arfken, George B., et al. *University Physics*. New York: Academic Press, 1984.

Becker, Friedrich. *Sternatlas*. Berlin: Ferd. Dümmlers Verlagsbuchhandlung, 1923.

Beyer, William H. *CRC Handbook in Mathematical Sciences*. Boca Raton, Fla.: CRC Press, 1987.

Brooks, Stewart M. *Going Metric*. San Diego, A. S. Barnes, 1976.

Chaikin, Andrew L., and Tom Hanks. *A Man on the Moon*. New York: Viking/Penguin Books, 1998.

Darton, Mike, and John O. E. Clark. *The Macmillan Dictionary of Measurement*. New York: Macmillan, 1994.

DIN Geschäftsbericht 1995/96. Berlin: Deutsches Institut für Normung e.V., 1996.

Donovan, Frank Robert. *Prepare Now for a Metric Future*. New York: Weybright and Talley, 1970.

Dubbel, Heinrich et al. *Handbook of Mechanical Engineering*. Berlin: Springer-Verlag, 1994.

Encyclopedia of Physics. New York: Macmillan, 1996.

Fermi, Laura. *My Life with Enrico Fermi*. Abuquerque: University of New Mexico Press, 1954.

Gröber, H., and S. Erk. *Die Grundgesetze der Wärmeübertragung*. Berlin: Verlag von Julius Springer, 1933.

Grun, Bernard. *The Timetables of History*. New York: Simon and Schuster, 1975.

Haberland, Gustav. *Elektrotechnische Lehrbücher*. Leipzig: Max Jänecke Verlagsbuchhandlung, 1941.

Haeder, W., and E. Gärtner. *Die gesetzlichen Einheiten in der Technik*. Berlin: Deutsches Institut für Normung e.V., 1980.

Klein, Herbert Arthur. *The Science of Measurement*. New York: Dover, 1988.

Kurtz, Max. *Handbook of Applied Mathematics for Engineers and Scientists*. New York: McGraw-Hill, 1991.

Landels, J. G. *Engineering in the Ancient World.* Berkeley: University of California Press, 1978.

Lide, David R. *CRC Handbook of Chemistry and Physics.* Boca Raton, Fla.: CRC Press, 1999.

Mache, Henrich. *Einführung in die Theorie der Wärme.* Berlin: de Gruyter, 1921.

McCoubrey, Arthur O. *Guide for the Use of the International System of Units—The Modernized Metric System.* Gaithersburg, Md.: National Institute of Standards and Technology, 1991.

"Quality Transmission Components." In *Handbook of Metric Gears.* New Hyde Park, N.Y.: Designatronics, 1993.

Richter, Hugo. *Aufgaben aus der technischen Thermodynamik.* Berlin: Springer-Verlag, 1953.

Terrien, J., and J. de Boer. *SI—Sistema International der Unidades.* Brasilia: Instituto Nacional de Pesos e Medidas, Cia. Grafica Lux, 1971.

Thomas, Oswald. *Astronomie, Tatsachen und Probleme.* Vienna: Verlag Das Bergland-Buch, 1933.

Tuma, Jan J. *Handbook of Physical Calculations.* New York: McGraw-Hill, 1976.

White, Michael. *Isaac Newton.* Reading, Mass.: Addison-Wesley, 1959.

Walker, Peter M. B., ed. *Cambridge Dictionary of Science and Technology.* Cambridge, U.K.: Cambridge University Press, 1976.

· Index ·

Gay-Lussac, Joseph Louis, 111
general theory of relativity, 81
geostationary satellite, 144, 172
"German Treatise," 9
Global Positioning System (GPS),
  172
gon, 43
grad, 42
grain, 87
gram, 142
Grand Coulee Dam, 120
gravimeter, 69
gravitational attraction, law of,
  158
gravitational constant, 72, 75
gravity, 68, 69
grease spot photometer, 88
Gudeu (Sumerian king), 24
guillotine, 137
Gunter, Edmund, 21

HAM operators, 83
harmonic progression, 14
heat, 99
Hebrew calendar, 53
Hefner candle, 87
helakim, 53
hemlock, 188
hertz (Hz), 139–40
Hertz, Heinrich, 83
hieroglyphic numerals, 2
Hipparchos, 95, 157
Hippocratic crescents, 13
hogshead, 162
Homer, 35
horsepower, 120
Hyades, 5
hydrothermal vents, 116

imperial gallon, 162
impulse, 78
induction, 132
infinite series, 15
internal combustion engine, 117
International System of Units, 63

International System of Weights
  and Measures (SI), ix, 193
inverse square law, 49
irrational numbers, 12
isothermal process, 116

Jefferson, Thomas, 26
jeroboam, 162
Joule, James Prescott, 168
Joule lunar crater, 167

kamal, 47
Kelvin of Largs, Lord, 102
Kelvin temperature scale, 102
Kennedy, John F., 194
Kepler, Johannes, 154
Kepler's third law, 175
khoes, 152
kilogram, 142
kilowatt·hour, 119
kinematic viscosity, 164
kite, 141
knot, 41

Lambert, Johann H., 12
latitude, 28
leading, 162
League of Nations' calendar, 56
Leibnitz, Gottfried W., 14
Leyden jar, 131
Liber Abaci, 8
light, 83; velocity of, 34
light-year, 169
Linde, Carl von, 102
Linné, Carl von (Carolus Linnaeus), 102
log, 36
logarithmic scales, 21
lumen, 93
luminance, 93
lunation, 54

magnum, 162
Mannheim, Amédée, 21
Marechal's calendar, 55

D0952325